electricity is measured in *ohms*. The amount of resistance provided by different materials varies widely.

For example, most metals offer little resistance to the passage of electric current. However, porcelain, wood, pottery, and some other substances have a very high resistance to the flow of electricity. In fact, these substances can be used as insulators against the passage of electric current.

What Are the Hazards of Electricity?

The primary hazards of electricity and its use are:

- Shock
- Burns
- Arc-Blast
- Explosions
- Fires

Shock

Electric currents travel in closed circuits through some kind of conducting material. You get a shock when some part of your body becomes part of an electric circuit. An electric current enters the body at one point and exits the body at another location. High-voltage shocks can cause serious injury (especially burns) or death.

You will get a shock if you touch:

- Both wires of an electric circuit.
- One wire of an energized circuit and ground.
- Part of a machine which is "hot" because it is contacting an energized wire and the ground.

Don't take any chances with electricity. One mistake can cost you your life. The severity of the shock a person receives depends on several factors:

- How much electric current flows through the body.

- What path the electric current takes through the body.
- How much time elapses while the body is part of the electric circuit.

What Happens to the Body?

The effects of an electric shock on the body can range from a tingle in the part touching the circuit to immediate cardiac arrest. A severe shock can cause more damage to the body than is readily visible.

Relatively small burn marks may be all that are visible on the outside. However, a severely shocked person can suffer internal bleeding and severe destruction of tissues, muscles, and nerves. Finally, a person receiving an electric shock may suffer broken bones or other injuries that occur from falling after receiving a shock.

The Case of Water

Water presents an interesting and potentially dangerous situation. In its pure state, water is a poor conductor of electricity. However, if even small amounts of impurities are present in the water (salt and acid in perspiration, for example), it becomes a ready electrical conductor.

Therefore, if water is present anywhere in the work environment or on your skin, be extra careful around any source of electricity. Carelessness with the combination of water and electricity could cost you your life.

Burns

Burns can result when a person touches electrical wiring or equipment that is improperly used or maintained. Typically, such burn injuries occur on the hands.

Arc-Blast

Arc-blasts occur when high-amperage currents jump from one conductor to another through air, generally during opening or closing circuits, or when static electricity is discharged. Fire may occur if the arcing takes place in an atmosphere that contains an explosive mixture.

Explosions

Explosions occur when electricity provides a source of ignition for an explosive mixture in the atmosphere. Ignition can be due to overheated conductors or equipment, or normal arcing (sparking) at switch contacts. OSHA standards, the National Electrical Code, and related safety standards have precise requirements for electrical systems and equipment used in hazardous atmospheres.

Fires

Electricity is one of the most common causes of fire both in the home and workplace. Defective or misused electrical equipment is a major cause, with high resistance connections being one of the primary sources of ignition. High resistance connections occur where wires are improperly spliced or connected to other components such as receptacle outlets and switches.

Heat develops in an electrical conductor from the flow of current. This heat raises the temperature of the conductor. As a result, resistance in the conductor increases, further raising the temperature. Thus, circuits conducting a high rate of current and generating more resistance than it can handle, may create enough heat to cause a fire.

Causes of Electrical Accidents

As a power source, electricity can create conditions resulting in bodily harm, property damage, or both. It is important for you to understand how to avoid electrical hazards when you work with electrical power tools, maintain electrical equipment, or install equipment for electrical operation.

Accidents and injuries in working with electricity are caused by one or a combination of the following factors:

- Unsafe equipment and/or installation.
- Unsafe workplaces caused by environmental factors.
- Unsafe work practices.

As an employee, you can definitely affect the last factor, and can be involved in reporting instances of the first two factors so they can be remedied.

Preventing Electrical Accidents

Protection from electrical hazards is one way to prevent accidents caused by electric current. Protective methods to control electrical hazards include:

- Insulation.
- Electrical protective devices.
- Guarding.
- Grounding.
- PPE.

Insulation

Insulators of glass, mica, rubber, or plastic are put on electrical conductors to protect you from electrical hazards. Before you begin to work on any piece of electrical equipment, take a look at the insulation (on electrical cords, for example) to be sure there are no exposed electrical wires.

OSHA has established specific standards for the insulation that covers electrical conductors. The insulation is required to be suitable for the voltage and conditions under which the item will be used (temperature, moisture level, fumes, etc.).

Electrical Protective Devices

Electrical protective devices, including fuses, circuit breakers, and ground-fault circuit-interrupters (GFCIs), are critically important to electrical safety. These devices interrupt current flow when it exceeds the capacity of the conductor and should be installed where necessary.

Current can exceed the capacity of the conductor when a motor is overloaded, for example, when you ask a 10 horsepower motor to do the work of a 12 horsepower motor, or when a fault occurs, as when insulation fails in a circuit.

When a circuit is overloaded, the insulation becomes brittle over time. Eventually, it may crack and the circuit fails, or faults.

Fault occurs in two ways. Most of the time a fault will occur between a conductor and an enclosure. This is called a ground fault. Infrequently, a fault will occur between two conductors. This is called a short circuit.

A device which prevents current from exceeding the conductor's capacity creates a weak link in the circuit. In the case of a fuse, the fuse is destroyed before another part of the system is destroyed. In the case of a circuit breaker, a set of contacts opens the circuit. Unlike a fuse, a circuit breaker can be reused by reclosing the contacts. Fuses and circuit breakers are designed to protect equipment and facilities, and in so doing, they also provide considerable protection against shock.

However the only electrical protective device whose sole purpose is to protect people is the ground-fault circuit-interrupter. The GFCI is not an overcurrent device. It senses an imbalance in current flow over the normal path and opens the circuit. GFCIs are usually installed on circuits that are operated near water.

Guarding

Any "live" parts of electrical equipment operating at 50 volts or more must be guarded to avoid accidental contact. This protection can be accomplished in several different ways. The machinery or equipment can be located:

- In a room, enclosure, or vault accessible only to qualified personnel.
- Behind substantial screens or partitions which prevent easy access.
- On a balcony, platform, or gallery area which is elevated and not accessible to unqualified/unauthorized persons.
- At least eight feet above the floor of the work area.

Any entrance to an area containing "live" parts of electrical equipment must be marked with conspicuous warning signs. These signs should forbid entrance except by qualified persons.

Grounding

Grounding is necessary to protect you from electrical shock, safeguard against fire, and protect against damage to electrical equipment. There are two kinds of grounding:

- Electrical circuit or system grounding, accomplished when one conductor of the circuit is intentionally connected to earth, protects the circuit should lightning strike or other high voltage contact occur. Grounding a system also stabilizes the voltage in the system so expected voltage levels are not exceeded under normal conditions.
- Electrical equipment grounding occurs when the equipment grounding conductor provides a path for dangerous fault current to return to the system ground at the supply source of the circuit should the insulation fail.

When a tool or other piece of electrical equipment is grounded, a low-resistance path is intentionally created to the earth. This path has enough current-carrying capacity to prevent any buildup of voltages in the equipment which could pose a hazard to an employee using the equipment. Therefore, never remove the ground prong from a plug because the equipment no longer protects you from short circuits. If you're touching an ungrounded tool, you will become the path of least resistance to the ground.

Grounding does not guarantee that an employee will never receive a shock, or be injured or killed by electricity in the workplace. However, this simple procedure will substantially reduce the likelihood of such accidents. Be sure any equipment you work on is properly grounded.

Personal Protective Equipment

If you work in an area where there are potential electrical hazards, your employer must provide you with protective equipment. You must use electrical protective equipment (see 29 CFR 1910.137) appropriate for the body parts that need protection and for the work to be done. Electrical protective equipment includes insulating blankets, matting, gloves, sleeves, overshoes, face protection, and hard hats among other equipment specially made to protect you from electricity.

Safe Work Practices for Handling Electricity

If your job requires you to work with electrical equipment, you need to have a healthy respect for the power of electricity. In general, you should be sure that any tools you use are in good repair, that you use good judgment when working near electrical lines, and that you use appropriate protective equipment. Remember—if you're not sure, don't touch.

Lockout/Tagout

A word to qualified employees about deenergizing electrical equipment before you do any repairs on it or make an inspection. (The chapter on Lockout/Tagout provides more details on this procedure.) Common sense dictates that electrical equipment be deenergized before working on it, when feasible. (Circumstances where it might be infeasible to deenergize circuitry or equipment before working on it would include hazardous location ventilation equipment or the testing of fire alarm systems, for example, that can only be performed when the system is energized. OSHA only allows qualified persons to perform this kind of work.)

Having electrical current unexpectedly present when you are working on a piece of equipment is no joke! Before any repair work or inspection of a piece of electrical equipment is begun by an authorized person, the current should be turned off at the switch box, and the switch padlocked in the OFF position.

The other step in this procedure is the tagging of the switch or controls of the machine or other equipment which is currently locked out of service. The tag should indicate which circuits or pieces of equipment are out of service.

Work at Working Safely

Safety should be foremost in your mind when working with electrical equipment. You face hazards from the tools themselves and the electricity that powers them. It's up to you to wear protective equipment whenever it's specified, use all safety procedures, and work with tools correctly. Never let overconfidence or complacency lead to taking unnecessary risks. If you're not sure—don't touch. The following general rules apply to every piece of electrical equipment you use:

- Be sure your electrical equipment is maintained properly. Regularly inspect tools, cords, grounds, and accessories. Make repairs only if you are authorized to do so. Otherwise, arrange to have equipment repaired or replaced immediately.

- Be sure you use safety features like three-prong plugs, double-insulated tools, and safety switches. Be sure machine guards are in place and that you always follow proper procedures.

- Install or repair equipment only if you're qualified and authorized to do so. A faulty job may cause a fire or seriously injure you or other workers.

- Keep electric cables and cords clean and free from kinks. Never carry equipment by its cords.

ELECTRICAL SAFETY

- Use extension cords only when flexibility is necessary:
    - **Never** use them as substitutes for fixed wiring.
    - **Never** run them through holes in walls, ceilings floors, doorways, or windows.

    - **Never** use them where they are concealed behind walls, ceilings, or floors.
- Don't touch water, damp surfaces, ungrounded metal, or any bare wires if you are not protected. Wear approved rubber gloves when working with live wires or ungrounded surfaces, and wear rubber-soled shoes or boots when working on damp or wet surfaces.
- Don't wear metal objects (rings, watches, etc.) when working with electricity. They might cause arcing.

ELECTRICAL SAFETY

- If you are working near overhead power lines of 50 kilo-Volts (kV) or less, you or any equipment you are using must not come any closer than 10 ft from the lines. Add 4 inches of distance for every 10 kV over 50 kV.

Good work habits soon become second nature. Treat electricity with the respect it deserves and it will serve you efficiently and safely.

EMERGENCY RESPONSE: DEALING SAFELY WITH CHEMICAL MISHAPS

In a world that increasingly relies on chemicals, the transportation, handling, and storing of those substances concern everyone. People are concerned because hazardous chemical spills and exposure affect individuals, families, communities, and the ecological health of the entire planet.

Improper handling or control of hazardous chemicals or waste can result in a severe threat to workers and to the general public when things get out of hand. Even extremely hazardous chemicals or waste, however, need not unnecessarily endanger human health if they are handled properly.

At the heart of safe handling of chemicals is good planning. That is why any organization that deals with hazardous chemicals will have:

- An emergency response team in place, **or**
- Made arrangements with an outside contractor to respond to chemical emergencies.

Although you do your best to avoid accidents, mishaps with toxic chemicals still occur. When a hazardous chemical spills in your plant, you need to know the correct steps to take in controlling and reducing risks to people's health and to the environment.

Where Are the Regulations?

The Occupational Safety and Health Administration (OSHA) has issued a special regulation dealing specifically with spills of chemicals. This regulation, called the Hazardous Waste Operations and Emergency Response (HAZWOPER) standard, is found at 29 CFR 1910.120.

Some employers covered by HAZWOPER opt instead to evacuate all employees in the event of a chemical emergency. These employers do not allow workers to assist in handling the spill and must have an emergency action plan (EAP) in place. Regulations covering EAPs are found at 29 CFR 1910.38(a).

Emergency Action Plans

You should be aware of certain procedures to protect yourself and others from injury during chemical and other emergencies. Your employer will conduct regular emergency drills so that you and your co-workers will know what to do and where to proceed during an emergency. You should be familiar with:

- How to report fires, hazardous chemical spills, and other emergencies.
- The route you are assigned to take during a building evacuation.
- Who to ask for more information.

Alarm Systems

Most employers use alarm systems to tell employees that they should evacuate an area or take a specific action. You must be able to recognize these alarms. In areas where production noise could prevent an alarm from being heard, flashing lights are often installed as a second, visual alarm. These alarm systems can generally also operate from auxiliary power sources so that they can operate even when the power goes out.

Emergency Shutdown of Equipment

If time permits before evacuation, turn off any equipment you are operating, such as forklifts or conveyors. Your employer may designate certain workers to shutdown critical facility systems, such as gas and electrical power, before evacuating the work area.

Evacuation

Under Emergency Action Plan requirements, your employer will develop emergency escape routes for the various locations in your facility. Floor plans or work-area maps clearly define emergency escape routes and are commonly used to convey this information.

Emergency evacuation procedures may also indicate shelter areas, such as locker rooms and cafeterias that are structurally sound, and the best routes to these areas. You might need to take shelter during a tornado, for example.

Your employer will designate certain employees to take a head count of all workers after evacuation and to inform emergency responders of any missing personnel.

Emergency Response Plans

Those employers who choose to handle chemical spills themselves must follow stringent requirements. Very specific training is necessary when preparing to handle accidental chemical releases. OSHA has set up a formal training schedule for emergency responders under HAZWOPER regulations, with training levels ranging from awareness training for first response to technical training for those with responsibility for solving problems associated with spill cleanup. **Under no circumstances does OSHA permit personnel without appropriate training to respond to a chemical spill emergency.**

The following information in this chapter outlines some considerations in handling a chemical spill emergency and **assumes that appropriate training has occurred**.

Your Role in the Event of a Spill

Whether it's a solid or a liquid spill, remember that you can be exposed to toxic dust or vapor without even knowing it. If you are properly trained, respond to a hazardous chemical spill with care and speed. However, if you have **not received training**, get to a safe place and sound an alarm according to established policy. At the least, a supervisor should be notified.

If a fire occurs, get out of the area where the spill occurred and call for help immediately. If you are trained in first aid, administer emergency first aid as necessary, but don't try to be a hero.

EMERGENCY RESPONSE

When a spill occurs it is vital to avoid panic. It is also important to avoid harm's way. Stay clear of the spill and notify your plant's emergency response team leader. He or she will assemble the rest of the team and take whatever emergency action is needed. Before the team arrives, there are still some things to think about:

KNOW potential hazards.

The material safety data sheet (MSDS) is one of your best sources of information on the chemicals used in your area. Have MSDSs at hand before responding to a spill.

KNOW about spill equipment and safety personnel.

Know where the emergency equipment is and how to use it or how to summon personnel who can use it properly. Know where the first aid equipment is located in your company.

KNOW the location of fire extinguishers.

Many chemicals are flammable or create a flammable product when mixed. In most cases of fire, you should get yourself and others out of the work area quickly and call for help. Use an appropriate fire extinguisher only if the fire is contained and is not rapidly spreading.

EMPLOYEE'S RECEIPT

I acknowledge receipt of Keller's Official OSHA Safety Handbook, which covers 23 different safety topics. These topics include:

    Confined Space Entry (1910.146)
    Electrical Safety (1910.303-.306)
    Emergency Response (1910.120)
    Ergonomics
    Fire Prevention (1910.38, .155-.165)
    First Aid and Bloodborne Pathogens
        (1910.151, .1030)
    Forklift Safety (1910.178)
    Hand Tools
    Hazard Communication (1910.1200)
    Lifting Techniques
    Lockout/Tagout (1910.147)
    Machine Guarding (1910.212)
    Personal Protective Equipment
        Eye Protection (1910.132, 133)
        Fall Protection
        Foot Protection (1910.132, .136)
        Hand Protection (1910.132, .138)
        Head Protection (1910.132, .135)
        Hearing Conservation (1910.95)
        Respiratory Protection (1910.132, .134)
    Slips, Trips and Falls (1910.21-.32)
    Violence in the Workplace
    Welding, Cutting, & Brazing
        (1910.251-.257)

_____   _____
Employee's Signature   Date

_____
Company

_____
Company Supervisor's Signature

> NOTE: This receipt shall be read and signed by the employee. A responsible company supervisor shall countersign the receipt and place it in the employee's training file.

Keller's Official OSHA Safety Handbook

4th Edition

©Copyright 1998

J. J. Keller & Associates, Inc.
3003 W. Breezewood Lane, P. O. Box 368
Neenah, Wisconsin 54957-0368
Phone: (920) 722-2848
www.jjkeller.com

United States laws and Federal regulations published as promulgated are in public domain. However, their compilation and arrangement along with other materials in this publication are subject to the copyright notice.

Library of Congress Catalog Card Number:
93-78349

ISBN 1-57943-611-0

Canadian Goods and Services Tax (GST)
Number: R123-317687

Printed in the U.S.A.

Fourth Edition, Second Printing, August 1999

All rights reserved. Neither the handbook nor any part thereof may be reproduced in any manner without written permission of the Publisher.

Due to the constantly changing nature of government regulations, it is impossible to guarantee absolute accuracy of the material contained herein. The Publishers and Editors, therefore, cannot assume any responsibility for omissions, errors, misprinting, or ambiguity contained within this publication and shall not be held liable in any degree for any loss or injury caused by such omission, error, misprinting or ambiguity presented in this publication.

This publication is designed to provide reasonably accurate and authoritative information in regard to the subject matter covered. It is sold with the understanding that the publisher is not engaged in rendering legal, accounting, or other professional service. If legal advice or other expert assistance is required, the services of a competent professional person should be sought.

The Editorial Staff is available to provide information generally associated with this publication to a normal and reasonable extent, and at the option of and as a courtesy of the Publisher.

TABLE OF CONTENTS

Introduction to the Worker

Confined Space Entry: How to Do it Safely (1910.146) 1

Electrical Safety: Don't Sell Yourself Short (1910.303–.399) 8

Emergency Response: Dealing Safely With Chemical Mishaps (1910.38(a), .120) 22

Ergonomics: Work That Fits People 33

Fire Prevention: Learn Not to Burn (1910.38(b), 155–.165) 45

First Aid and Bloodborne Pathogens: Protecting Yourself While Giving Assistance (1910.151, .1030) 58

Forklift Safety: Tips for Designated Drivers (1910.178) 69

Hand Tools: The Right Tool for the Job 81

Hazard Communication: The Right to Know Law (1910.1200) 87

Lifting Techniques: Avoiding Back Injuries 97

Lockout/Tagout: The Control of Hazardous Energy (1910.147) 106

Machine Guarding: Working Safely With Machines (1910.212) 117

Personal Protective Equipment 135
    Eye Protection: Seeing Is Believing (1910.132, .133) . 141
    Fall Protection: Knowledge Is Standard Equipment ... 153
    Foot Protection: Keeping on Your Toes (1910.132, .136) 166
    Hand Protection: Let Your Fingers Do the Working (1910.132, .138) 177
    Head Protection (1910.132, .135) 186
    Hearing Conservation: Now Hear This (1910.95) 189
    Respiratory Protection: Choosing and Using Respirators (1910.132, .134) 200

Slips, Trips and Falls: On the Job Safety Basics 211

Violence in the Workplace 224

Welding, Cutting and Brazing: Avoid the Triple Threat 231

Chapter Reviews

Introduction to the Worker

Your employer cares about your safety. Keller's Official OSHA Safety Handbook is just part of its effort to teach you about the hazards you face on the job every day. This handbook will be part of the overall safety program your company has set up to train and protect its workers.

Occupational Safety and Health Administration (OSHA) regulations often require training on particular topics, such as occupational noise exposure or personal protective equipment (PPE). This handbook covers topics important to your safety and health on the job, including subjects like hazard communication and fire prevention. It serves as an aid to your supervisor or instructor in training you on your facility's safety procedures. You need to learn what to do to be safe at work, and your employer needs to teach you what to do. This handbook helps your employer teach you about many topics vital to workplace safety and health.

OSHA regulations for general industry have been issued in the Code of Federal Regulations, Title 29, Part 1910. Individual regulations will be identified throughout the handbook as 29 CFR 1910.(section number).

This book is yours to keep. Use it to follow along with your instructor during training sessions, then keep it to use as a handy reference after you finish your safety training program. You can look up safety and health facts and answer safety questions that occur while you're working. The handbook can be kept in your locker or toolbox, to keep safety information at hand whenever you need it.

Together with your employer, you can make every workday a safe one by following the guidelines in Keller's Official OSHA Safety Handbook.

CONFINED SPACE ENTRY: HOW TO DO IT SAFELY

What Is a Confined Space?

Thousands of workers are exposed to possible death or injury in what are referred to as "confined spaces." The Occupational Safety and Health Administration (OSHA) estimates that about 224,000 establishments employing 7.2 million workers have confined spaces on their property or work areas. Over 5,000 injuries occur in confined spaces each year.

The OSHA permit-required confined space standard defines a confined space as a space that is large enough for an employee to enter, has restricted means of entry or exit, and is not designed for continuous employee occupancy. A permit-required confined space (permit space) is a confined space that presents or has the potential for hazards related to atmospheric conditions (toxic, flammable, asphyxiating), engulfment, configuration, or any other recognized serious hazard.

Where Are the Regulations?

Regulations governing entry to confined spaces are specified by OSHA. You can find these regulations at 29 CFR 1910.146.

Examples of confined spaces include ship compartments, missile fuel tanks, vats, silos, sewers, tunnels, and vaults. Although these environments are often dangerous, you might have to work in them to inspect equipment, hardware, or structural elements, or to clean, repair, or maintain them.

CONFINED

Confined Space Hazards

What makes a confined space hazardous? Dangerous vapors and gases can accumulate in these spaces. Fires, explosions and physical hazards can also injure or kill an unprotected worker. Let's look further at some of the dangers you might encounter in a confined space.

Physical Hazards

Physical hazards may result from mechanical equipment or moving parts like agitators, blenders, and stirrers. Dangers may also be present from gases, liquids, or fluids entering the space from connecting pipes. Before entering a permit space, all mechanical equipment must be locked out/tagged out. All lines containing hazardous materials such as steam, gases, or coolants should also be shut off.

Other physical hazards include heat and sound. Temperatures can build up quickly in a permit space and cause exhaustion or dizziness. Sounds may reverberate and make it hard to hear important directions or warnings.

Oxygen Deficiency

Most permit space accidents are related to atmospheric conditions inside the space or to the failure to continuously monitor hazards. In general, the primary hazard associated with confined spaces is oxygen deficiency.

Normal air contains 20.8 percent oxygen by volume. The minimum safe level as indicated by OSHA is 19.5 percent; OSHA defines the maximum safe level as 23.5 percent. At 16 percent you will feel disoriented, and between 8 and 12 percent, you could become unconscious. Oxygen is reduced in a space either by displacement or consumption. Oxygen is displaced by other gases such as argon, nitrogen, or methane. Consumption may be caused by chemical reactions such as rusting, rotting, fermentation, or burning of flammable substances.

Combustibility

Fire and explosion are serious dangers in a permit space. Flammable and combustible gases or vapors may be present from previous cargoes, tank coatings, preservatives, and welding gases. These built-up vapors and gases can be ignited by faulty electrical equipment, static electricity, sparks from welding, or cigarettes.

Toxic Air Contaminants

These contaminants occur from material previously stored in the tank or as a result of the use of coatings, cleaning solvents, or preservatives. Unfortunately, you won't be able to see or smell most toxics, but they present two types of risk in a permit space.

First, they can irritate your respiratory or nervous system. Second, some toxic chemicals can cut off your oxygen supply or get into your lungs and asphyxiate you.

Working in Confined Spaces

Once you recognize a hazard, there are several important steps to follow. You should plan carefully before entering the space at all. Then you should test the air before entry and periodically as you work. If there are any hazards in the space, it is a permit-required confined space and you may enter only if you're following your facility's confined space permit program. There is an alternate procedure for confined spaces when the only hazard is a hazardous atmosphere that can be made safe for entry through continuous ventilation. These spaces don't need permits.

Finally, you should have a rescue plan in case of an emergency. Work in and around permit spaces must be viewed as a total process. Safe entry is only the beginning. You want to enter, complete your work, and exit safely.

Before You Enter

Entry permits must be used for entry into a confined space that presents or has the potential for hazards related to atmospheric conditions or any recognized serious hazard. In such cases, only a worker with a written permit should be allowed to enter a permit-required confined space. In situations where you must use the permits, get one from your supervisor and post it outside the permit space to warn others that you are inside.

Control any Hazardous Energy

Use locks and tags to prevent accidental startup of equipment while someone is working in the permit space. Cut off steam, water, gas, or power lines that enter the space. Use only safe, grounded, explosion-proof equipment and fans.

Test the Air

Use special instruments for testing the levels of oxygen, combustibility, and toxicity in confined spaces. A common cause of injury and death in permit space accidents has been failure to test before entry, and during work procedures, for dangerous air contaminants and safe oxygen levels.

Test for oxygen and combustibility before you open the space by probing with test instruments near the entry.

Once the space is opened, test the air from top to bottom. Some gases like propane and butane are heavy, and they will sink to the bottom of the space. Light gases like methane will rise to the top. So you need to be sure to check all levels.

After you are sure that the oxygen level is adequate and there is nothing combustible in the space, test for toxicity. Notify your supervisor if pretests find hazards that you can't protect against adequately. Follow-up testing may be periodic or continuous, depending on specific conditions. Because the work being done within a permit space may also change air quality, continuous testing may be needed to ensure worker safety.

Purge and Ventilate the Space

Some confined spaces may contain water, sediment, hazardous atmospheres, or other unwanted substances. These substances generally must be purged, that is, pumped out or otherwise removed, before entry.

Use ventilating equipment where possible. Ventilation should maintain an oxygen level between 19.5 percent and 23.5 percent. It also should keep toxic gases and vapors to within accepted levels prescribed by OSHA.

Portable self-contained breathing devices should be used where the entrance is large enough, unless the atmosphere is ventilated or has been determined to have no potential atmospheric hazards. If the entrance to a potentially hazardous atmosphere is too small, an airline mask should be used with supplied air. If there is any danger of disconnecting the airline, a 5 to 15-minute escape respirator containing breathing air should be carried. You might also need eye and hearing protection and protective clothing.

Rescue Procedures

OSHA estimates that about 54 workers die each year in permit spaces. Almost two-thirds of those deaths result from people attempting a rescue. When workers enter a permit space, at least one person must remain outside to summon help or offer assistance.

When the employer designates this attendant to perform rescue procedures in addition to monitoring the safety of the confined space entrant, he or she should be equipped with the necessary PPE and rescue equipment and trained in first aid and cardiopulmonary resuscitation (CPR). He or she should maintain constant communication with those inside the space either visually, by radio, or by field telephone. If a situation arises that requires emergency entry, the attendant should not enter until additional help arrives.

A rope tied around a worker's waist is not an acceptable rescue method because it does not allow a single attendant to pull an injured worker out of a space. A full body harness and lifeline is a better approach because it can be attached to a block and tackle which a single rescuer/attendant can operate.

The entrant must be trained to recognize hazardous situations so he or she knows when to exit a confined space before a rescue is needed.

Work at Working Safely

You should always follow safety procedures and your company's permit space program, and use protective equipment made available by your employer. Keep the following safety tips in mind:

1. Use prescribed personal protective and respiratory equipment at all times.
2. Test the air inside the permit space for safe oxygen levels and for flammable, explosive, and toxic vapors and gases before entry. If necessary, test again while work is in progress to ensure continued worker safety.
3. Always use spark-proof tools and explosion proof fans, lights or air movers when working in permit spaces.
4. Have trained, well-equipped workers standing by to rescue anyone who gets into trouble while working in a permit space.

If you follow these rules carefully, you will be able to work safely even in permit-required confined spaces.

ELECTRICAL SAFETY: DON'T SELL YOURSELF SHORT

Electricity is such an integral part of our lives at home and in the workplace that we can tend to take its power for granted. But here's a sobering fact.

The Bureau of Labor Statistics (BLS) reported that during 1996, 279 (or about five percent) of work-related deaths in the private sector resulted from electrocution. Don't become another tragic statistic. Electrical accidents in the workplace can, for the most part, be avoided if you use safe electrical equipment and work practices.

Where Are the Regulations?

The Occupational Safety and Health Administration has devoted an entire section (Subpart S) of its regulations to rules governing electrical work. These regulations can be found in 29 CFR 1910.301-.399. In addition, you may need to comply with regulations relating to personal protective equipment, found in Subpart I of the General Industry regulations.

Your employer must train you in safe work practices for working with electrical equipment. Section 1910.332 specifically covers training related to your job. The training rules distinguish between workers who work on or near exposed energized parts and those who do not. Even if you are not qualified to work on electrically energized equipment, you must know the specific safety practices which apply to your job.

Are You Qualified to Work With Electricity?

OSHA has designated two categories of workers who face a risk of electric shock that is not reduced to a safe level by using electrical protective devices and following safe work practices. These categories identify workers who are:

- Unqualified, that is, those workers who are trained in and familiar with safety-related work practices required in Subpart S.

- Qualified, that is, those workers trained on avoiding the hazards of working on or near exposed live parts, in addition to the training required for unqualified workers. Qualified workers would have been trained: to distinguish exposed live parts from other parts of electric equipment, in methods to determine nominal voltage of exposed live parts, and to know the proper clearance distances.

This chapter focuses on information needed by workers who are unqualified.

How Does Electricity Work?

To handle electricity safely, including working with electrical equipment, you need to understand how electricity acts, how it can be approached, the hazards it presents, and how those hazards can be controlled.

Basically, there are two kinds of electricity:

- Static (stationary).
- Dynamic (moving).

This chapter is about dynamic electricity because that is the kind commonly put to use. Dynamic electricity is the flow of electrons through a conductor. An electron is a tiny particle of matter that orbits around the nucleus of an atom. Electrons of some atoms are easily moved out of their orbits. This ability of electrons to move or flow is the basis of electrical current.

When you activate a switch to turn on an electric machine or tool, you allow current to flow from the generating source through conductors (usually wires) to the area of demand.

SIMPLE CIRCUIT
Electron Flow

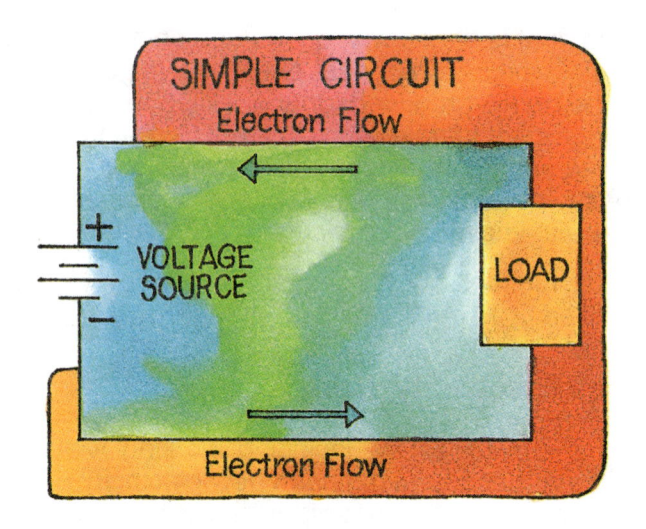

VOLTAGE SOURCE

LOAD

Electron Flow

A complete circuit is necessary for the controlled flow of electrons along a conductor. A complete circuit is made up of a source of electricity, a conductor, and a consuming device (load).

Volts = Current × Resistance (or V=IR)

The title of this section, Volts = Current × Resistance, is an equation known as Ohm's Law. The factors discussed below relate to one another as described by this equation. This relationship makes it possible to change the qualities of an electrical current but keep an equivalent amount of power.

A force or pressure must be present before water will flow through a pipeline. Similarly, electrons flow through a conductor because *electomotive force* (EMF) is exerted. The unit of measure for EMF is the *volt*.

For electrons to move in a particular direction, a potential difference must exist between two points of the EMF source. For example a battery has positive and negative poles.

The continuous movement of electrons past a given point is known as *current*. It is measured in *amperes*. The movement of electrons along a conductor meets with some opposition. This opposition is known as *resistance*. Resistance to the flow of

KNOW the location of emergency exits.

Make sure you know how to get out of your work area or the building quickly if necessary. Clear work area obstructions like chairs and lab carts so that you and co-workers are not endangered because of a slow exit from the scene of a toxic spill.

KNOW first aid or where to get it.

Know the general first-aid rules and what the MSDSs say about first aid for the particular substances you work with.

Cooperate with the emergency response team when it arrives by passing on any information you've gathered.

Don't forget to use the buddy system. Members of the emergency response team should never enter a chemical emergency situation alone.

The Emergency Response Team

Members of an emergency response team need to be both physically and emotionally fit to handle stressful situations. Those selected for the team in your facility will probably have a complete physical exam as well as psychological or personality testing.

Once the team has been selected, the leader will explain the responsibilities of its members. Teamwork is the key to the success of an emergency response operation.

Your team will go through training sessions throughout your plant as it becomes familiar with a written work plan that addresses personnel roles, lines of authority, communications, emergency recognition and prevention, places of refuge, site security, evacuation routes, and emergency medical care.

The object is for the team to perform its job with minimal risks. Team members must know the limits for themselves and their equipment. They need to know how to safely avoid or escape emergency situations.

Evacuation During Spill Response

Response team leaders will consider evacuation as soon as a spill is evaluated. The decision to evacuate will depend on the chemical and the hazards involved.

Evacuation can be as simple as clearing people from the immediate area of the spill or as complicated as total evacuation of the plant, surrounding area, and community. In total evacuations involving the community, proper public safety personnel must be involved.

PPE for Chemical Handling

The hazardous chemicals used at your plant or work site determine the level of personal protective equipment (PPE) and clothing. You will receive necessary instruction and training on PPE and clothing.

The Environmental Protection Agency has set four levels of personal protection, A through D, with A providing most protection:

- Level A PPE consists of the most protective type of respirator; a fully encapsulating, chemical-type resistant suit; inner and outer rubber gloves; rubber boots and steel toe safety boots; and a two-way radio. A hard hat may also be needed.
- Level B consists of the same type of respiratory protection as Level A; disposable, hooded chemical-resistant coveralls; rubber boots and steel toe safety boots; inner and outer rubber gloves; and a two-way radio. A hard hat may also be needed.
- Level C includes the same personal protective clothing mentioned for Level B, but with an air-purifying respirator.
- Level D consists of a basic work uniform that provides minimal chemical protection. Level D may be used where contamination is at nuisance levels only. Level D may include coveralls, eye protection, safety shoes, gloves, etc.

EMERGENCY RESPONSE

Spill Carts and Spill Control Stations

To deal with spills and other accidents, spill carts and spill control stations are frequently used. Know the location of the carts and stations. Keep them accessible and well stocked. Again, you will receive instruction and training as your job requires.

Typical items found in spill carts (or wagons) and spill control stations are:

- Pillows, pads, or other materials designed to collect, neutralize, absorb, or suppress hazardous liquids while picking up a spill.
- Patch and plug kits.
- Brooms, mops, scrapers, squeegees, and buckets.
- Both acid and base neutralizers.
- Temporary warning labels.
- Tapes and barricades.
- Coveralls, goggles, and gloves.
- Salvage drums and waste containers.

You need to make sure you have the supplies and equipment you need for your work area—and that you have adequate quantities.

Decontamination Procedures

A decontamination plan needs to be part of every plant's emergency response effort so that personal protective equipment and clothing are selected with both response to the emergency and decontamination in mind. Planning ahead makes sense. Dispose of chemically contaminated waste properly.

Emergency Follow-up Is Essential

Following an emergency, OSHA must be notified if the incident resulted in any fatalities or if three or more persons were hospitalized. If the spill is significant, the National Response Center must be notified as well. The Environmental Protection Agency regulates chemical releases to the environment.

The final activity of the emergency response team following any emergency is to review and evaluate all aspects of what happened and what may happen as a result. An account of the incident must be accurate, authentic, and complete, so be prepared to cooperate.

Work at Working Safely

Help to make your workplace safe and healthy by following safety guidelines, especially as they apply to handling hazardous chemicals. Remember, the best accident and spill control is accident and spill prevention. But in the event of a spill:

1. Tell others a spill has occurred and help them get to safety.
2. Report the spill to management.
3. Don't ignore a spill. It can produce harmful or deadly vapors.
4. Don't endanger your own life. If fire or explosion seem imminent, get out of the work area.

ERGONOMICS: WORK THAT FITS PEOPLE

What Is Ergonomics?

Ergonomics is a discipline that involves arranging the environment to fit the person in it. An example of good ergonomic design can be as simple as a well-designed knife, screwdriver, or pliers, or as elaborate as equipment on a production line designed to accommodate operators.

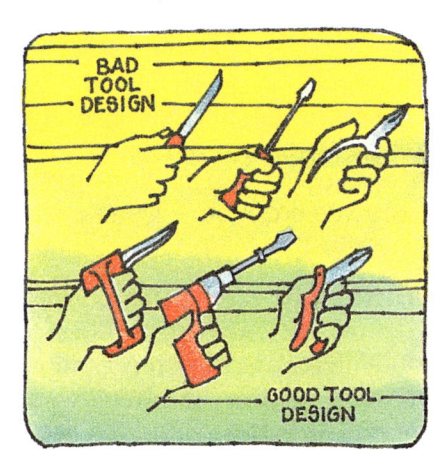

Following ergonomic principles helps reduce stress and eliminate many potential injuries and disorders associated with overuse of muscles, bad posture, and repeated physical tasks. The objective of ergonomics is to reduce worker stress and injury through design of tasks, work stations, controls, displays, safety devices, tools, lighting, and equipment.

Why Is Ergonomics a Concern in the Workplace?

Technological advances resulting in more specialized tasks, higher assembly line speeds, and increased repetition can be related to chronic or acute injury. This is because workers' hands, wrists, arms, shoulders, backs, and legs may be subjected to thousands of repetitive twisting, forceful, or flexing motions during a typical workday.

Some jobs expose workers to excessive vibration and noise, eye strain, repeated movements, and heavy lifting. Machines, tools, and the work environment may be designed so that stress is placed on workers' tendons, muscles, and nerves. In addition, workplace temperature extremes may aggravate or increase ergonomic stress. Recognizing ergonomic hazards in the workplace is the first step in improving worker protection.

Where Are the Regulations?

The Occupational Safety and Health Administration (OSHA) has researched the feasibility of regulating ergonomics hazards, but has not completed its study. OSHA plans to continue to investigate ergonomic hazards and cite employers using Section 5(a)(1) (the General Duty Clause) of the OSH Act as the basis for issuing citations. Even without being forced by a regulation, some employers are finding that taking relatively simple steps can resolve ergonomic issues.

Manager Commitment and Employee Involvement

Commitment and involvement are complementary and essential elements of a sound safety and health program. Commitment by management provides the organizational resources and motivating forces necessary to deal effectively with ergonomic hazards.

The implementation of an effective ergonomics program includes a commitment by the employer to provide the visible involvement of top management, so that all employees, from management to line workers, fully understand that management has a serious commitment to the program.

Employee involvement and feedback, through clearly established procedures, are likewise essential, both to identify existing and potential hazards and to develop and implement an effective way to abate such hazards. Employees' intimate knowledge of the jobs they perform and the special concerns they bring to the job give them a unique perspective which can be used to make the program more effective.

Disorders/Injuries Related to Ergonomic Hazards

A variety of disorders and illnesses related to muscles and bones are caused by ergonomic stressors. These disorders, called musculo-skeletal disorders, involve the muscles, tendons, ligaments, nerves, joints, bones, or supporting body tissue. Injuries include disorders, of the back, the neck, the upper or lower extremities, or the shoulders, and involve strains, sprains, or tissue inflammation.

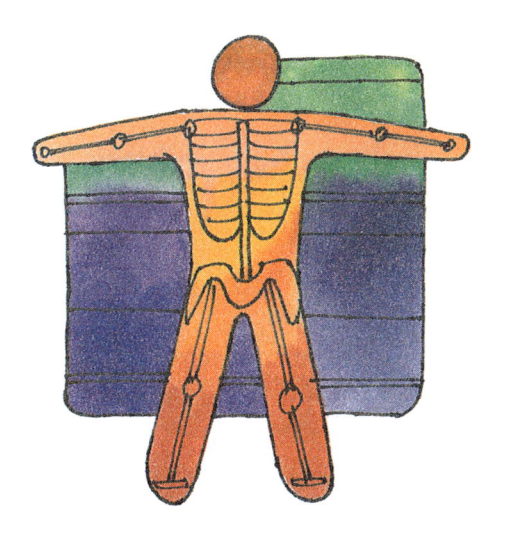

Worker/workplace interactions which cause musculo-skeletal disorders include heavy lifting, constant twisting, and repeated motions. In addition, physical characteristics of the worker vary from human to human, including size, endurance, range of motion, strength, gender, and other factors. When the job demand exceeds the physical characteristics of the worker, an injury is likely to result.

Cumulative Trauma Disorders

Cumulative trauma disorders (CTDs) are disorders of the musculo-skeletal and nervous systems which are caused or made worse by repetitive motions, forceful exertions, vibration, hard and sharp edges, or sustained or awkward postures.

CTDs can affect nearly all tissues, the nerves, tendons, tendon sheaths, and muscles, with the upper extremities being the most frequently affected. These injuries develop gradually and result from repeated forceful actions, such as twisting and bending of the hands, arms, and wrists.

CTDs in the workplace are often tendon disorders, which may occur at or near the joints where the tendons rub against ligaments and bones. The most frequently noted symptoms of tendon disorders are a dull aching sensation over the tendon, discomfort with specific movements, and tenderness to the touch. Recovery is usually slow, and the condition may easily become chronic if the cause is not eliminated. Tendon disorders can arise from a wide variety of occupational motions and actions.

Vibration along with prolonged sitting may also result in degenerative changes in the spine. For example, drivers of tractors, trucks, buses, construction machinery, and other heavy equipment may suffer from low back pain, and permanent abdominal, spinal, and bone damage.

Carpal Tunnel Syndrome (CTS) is a specific CTD affecting the hands and wrists, and has probably received more attention in recent years than any other CTD.

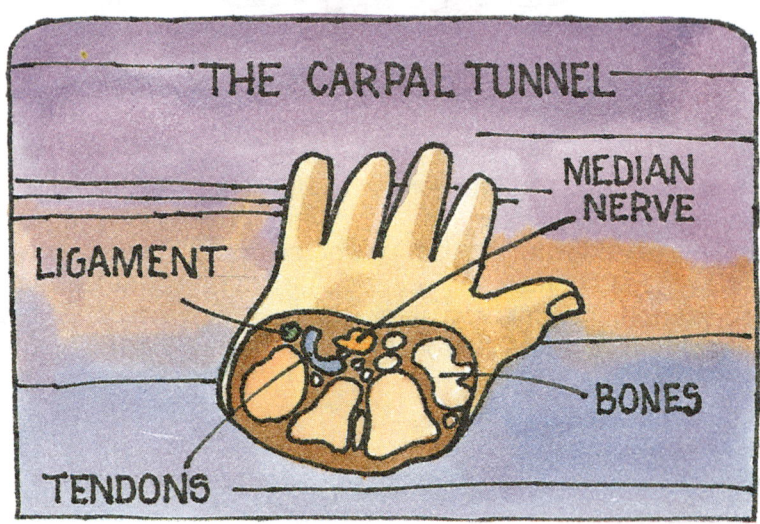

THE CARPAL TUNNEL

MEDIAN NERVE

LIGAMENT

BONES

TENDONS

The pressure of repetitive motion causes tingling, numbness, or severe pain in the wrist and hand. The pressure also results in a lack of strength in the hand and an inability to make a fist, hold objects, or perform other manual tasks. If the pressure continues, it can damage the nerve, causing permanent loss of sensation and even partial paralysis.

CTS develops in the hands and wrists when repetitive or forceful manual tasks are performed over a period of time. For example, the meatpacking industry is considered hazardous because workers can make as many as 10,000 repetitive motions per day in assembly line processes, such as deboning meats, with no variation in motion. CTS is common among meat and poultry workers, carpenters, garment workers, shoe makers, assemblers, packers, product inspectors, machine operators, keypunch operators, office workers, and cashiers.

Employees are often unaware of the causes of CTS and what to do about them. Often they do not associate their pain with their work because symptoms may only occur during evening or off-duty hours.

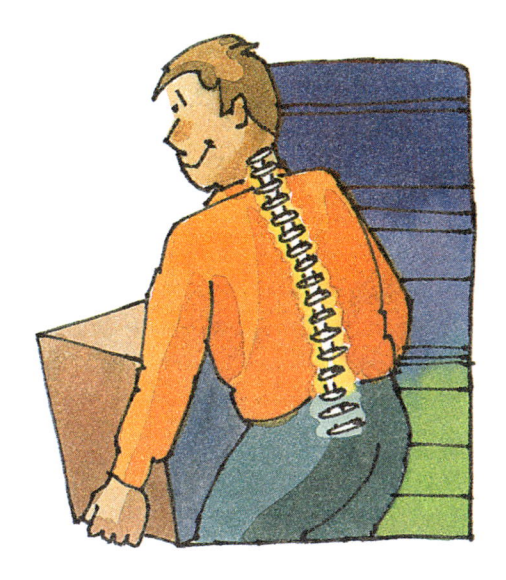

Back Disorders

Pulled or strained muscles, ligaments, tendons, and disks are perhaps the most common back problems and may occur in almost half of the work force at least once during their lifetime.

Most workplace back disorders result from chronic, or long-term injury to the back, rather than from one specific incident.

Back disorders are frequently caused by:

- Excessive or repetitive twisting, bending and reaching.
- Carrying, moving, or lifting loads that are too heavy or too big. (See chapter on Lifting Techniques for more detail.)

- Staying in one position for too long.
- Poor physical condition.
- Poor posture.

Prolonged sitting stresses the body, particularly the lower back and the thighs, and may cause the lower back (lumbar) region to bow outward if there is inadequate support. This abnormal curvature (called kyphosis) can lead to painful lower back problems, a common complaint among office workers. Other factors which are contributors to back injuries include:

- The natural degeneration of the back due to aging.
- Inactivity, both at work and at home.
- Seasonal activity undertaken without prior physical conditioning.
- Stress.
- Vibration.

According to one source, truck drivers have the highest incidence of back injuries, most likely due to prolonged sitting in one position and prolonged exposure to vibration. Many other workers, however, are at risk of developing back disorders, including those in industries such as manufacturing, construction, transportation, shipbuilding, and hospital care.

Contributing Factors

Studies have shown that work-related accidents increase with both higher and lower workplace temperatures. Where temperature extremes require workers to use more force in performing their jobs, ergonomic stress and risk of ergonomic disorders may also increase.

Cold temperatures can affect a worker's coordination and manual dexterity, thus requiring more effort and additional manual force to perform the same task or to maintain productivity levels. Likewise, hot and humid conditions may cause excessive fatigue or reduce the employee's work capacity, resulting in increased ergonomic stress. Such conditions may also require a worker to apply more force or effort in gripping hand tools or in using other equipment, which may further increase ergonomic stress and the risk of ergonomic disorders.

One way your employer may uncover ergonomic hazards is by doing a worksite analysis. Possible risk factors for cumulative trauma and back disorders that may be identified if you do a worksite analysis include:

- Regular repetitive tasks.
- Awkward postures.
- Forceful exertions.

- Temperature extremes.
- Inappropriate hand tools.
- Restrictive workstations.
- Vibration from power tools.
- Poor body mechanics.
- Lifting heavy or awkward objects.

The combined effect of several risk factors often results in the onset of CTDs. The ergonomic approach is to make things better than they were before; even small improvements in reducing or eliminating some, if not all, risk factors within a problem area will reduce the risk and the level of physical stress for the worker.

Hazard Prevention and Control

Ergonomic hazards are prevented primarily by the effective design of a job or jobsite and the tools or equipment used in that job. Based on information gathered in the worksite analysis, procedures can be established to correct or control ergonomic hazards using the following methods.

Engineering controls

The primary focus of fixing ergonomic problems is to make the job fit you, not force you to fit the job. Engineering controls can accomplish this by ergonomically designing work stations and tools or equipment.

Your work station should accommodate the full range of movements you need to do your job. It should let you assume several different but equally healthful and safe postures that still let you do your job:

- You should have enough space for your knees and feet.
- You should be able to adjust the height of worktables and chairs so that you have proper back and leg support. (You can use seat cushions to compensate for height variation when chairs or stools are not adjustable.)
- You should be able to easily reach machine controls.

When redesigning a work station, your supervisor will pay attention to activities that require you to exert force, the reaching height and distance needed for you to perform tasks, and the force requirements. Other factors he or she will look at include hard or sharp edges, contact with hot work surfaces, proper seating, work piece orientation, and layout of the workstation.

Other engineering adaptations can be made to tools and tool handles. Select ergonomically designed tools to improve productivity and prevent CTDs.

Because improper use and poor design of hand-held tools can cause damage to your hand and arm, several factors should be considered when selecting a tool, including:

- Position in which it will be used.
- Vibration.
- Grip strength required.
- Awkwardness.
- Force required.
- Repetitive motion involved.
- Handle sizes and their adequacy for gloved hands.

- Accommodation of both right- and left-handed workers.
- Balance.
- Center of gravity
- Weight of the tool.
- Whether or not the tool is appropriate for the job.

Administrative Controls

Administrative controls are used to reduce the duration, frequency, and severity of exposure to ergonomic stressors. Once ergonomic problems have been identified, your supervisor may opt to train you on ways to avoid injury at your job, reduce the number of repetitions you make, provide rest breaks, or begin cross-training so that you may rotate jobs with co-workers.

Medical Management

Your employer may institute a medical management program for cumulative trauma disorders to identify and treat the symptoms early. You need to be able to recognize the signs and symptoms of CTD so that you will report them and start treatment if necessary, have your job evaluated to avoid further irritation, and possibly go on a modified work schedule during the treatment period. You can assist in this process by seeking help before injuries become more severe, and in helping your supervisor understand exactly how you do your job.

Work Practice Controls

Key elements of a good work practice program include instruction in proper work techniques, employee training and conditioning, regular monitoring, feedback, adjustments, modification, and maintenance.

For example, after you are trained in a particular work activity, such as proper lifting or proper tool handling, your supervisor should monitor you to make sure you continue to use the proper techniques. Improper practices should be corrected to prevent injury.

Work at Working Safely

You need to be aware of ergonomics and the causes of musculo-skeletal and cumulative trauma disorders.

1. Cooperate with your employer in making ergonomically designed changes in the workplace. Ergonomic changes to design and layout will lessen the chance of injury to you.
2. Be aware of the signs and symptoms that may indicate a problem or injury caused by a poorly designed workplace.
3. Participate in any hazard control activities your employer initiates.
4. Become aware of job-specific techniques you can use to alleviate ergonomic problems.
5. See a doctor about any CTD-related injuries. A doctor can provide proper diagnosis and suggest possible solutions to avoid future workplace injury.

FIRE PREVENTION: LEARN NOT TO BURN

The best defense against a fire is to prevent a fire from starting in the first place. Although many products stored in a warehouse or work area are not flammable, some packaging types commonly used today, such as cardboard, excelsior, foam compositions, and paper packing are definite fire hazards. In addition, some of the chemicals you work with may be able to start or feed a fire.

You need to know what to do to keep fires from starting, as well as how to deal with the emergency of an accidental fire. Because of the deadly danger of fire, it's to your benefit to know how to size up a fire and how to respond in a fire emergency.

Where Are the Regulations?

The Occupational Safety and Health Administration (OSHA) regulates several aspects of fire prevention and response. Emergency planning, fire prevention plans, and evacuation are addressed in 29 CFR 1910.38. In addition, the provisions for fire extinguishers and other protection are addressed at 29 CFR 1910.157.

What Kind of Fire Is it?

The National Fire Protection Association (NFPA) has classified four general types of fires, based on the combustible materials involved and the kind of extinguisher needed to put them out. The four fire classifications are A, B, C, and D. Each classification has a special symbol and color identification.

General Classes of Fires

Class A. This type of fire is the most common. The combustible materials are wood, cloth, paper, rubber, and plastics. The common extinguishing agent is water, but dry chemicals are also effective. Do not use carbon dioxide extinguishers and those using sodium or potassium bicarbonate chemicals on these fires.

Class B. Flammable liquids, gases, and greases create class B fires. The most common extinguisher to use is dry chemical. Also, foam and carbon dioxide extinguishers can be used.

Class C. Because class C fires are electrical fires, use a non-conducting agent to put them out, for example, carbon dioxide and dry chemical extinguishers. Never use foam or water-type extinguishers on these fires.

Class D. Fires arising from combustible metals, such as magnesium, titanium, zirconium, and sodium are categorized as class D fires. These fires require specialized techniques to extinguish them. None of the common extinguishers should be used since they can increase the intensity of the fire by adding an additional chemical reaction. Use dry powder extinguishers specific for the metal hazard present on these fires.

There are only two dry chemical extinguishers that can be used on A, B, and C fires, and those are multi-purpose ABC extinguishers, either stored pressure or cartridge operated.

Multi-purpose extinguishers (ABC) will handle all A, B, and C fires. **All fire extinguishers are labeled with either ABC, or A, or B, or C, so be sure to read the label.**

Housekeeping to Prevent Fires

The importance of good housekeeping ties in closely with fire prevention. If you allow debris or flammable material to accumulate, the risk of starting a fire increases. There is always the possibility that fire may break out by accident. Fire prevention is part of everyone's job. Everyone must help to keep the work area clutter-free and safe from other fire hazards, such as improperly used or stored chemicals.

You also need to know what to do in the case of a fire emergency. Your employer has a fire prevention plan spelling out everyone's roles; you should know the actions you are expected to take in the event of a fire.

ORDINARY

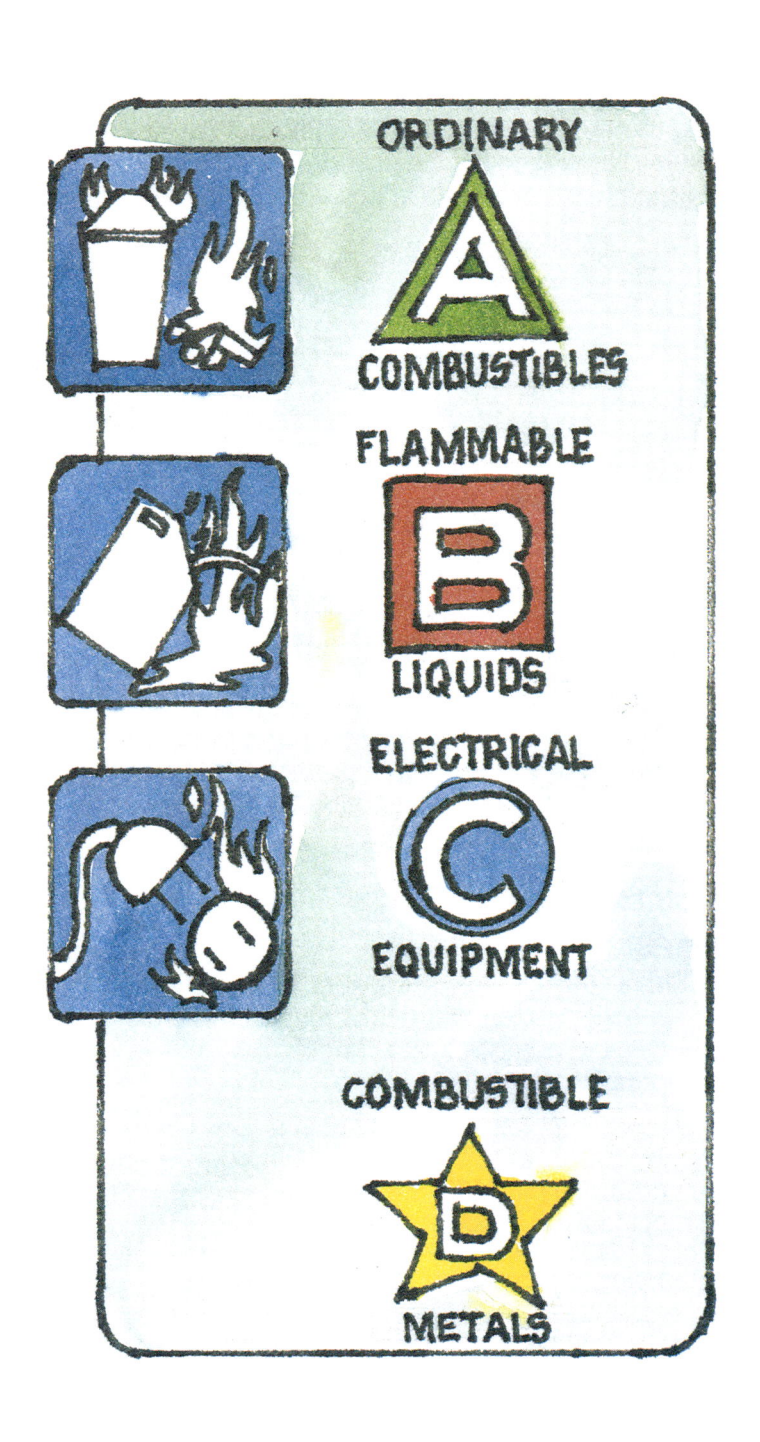

A

COMBUSTIBLES

FLAMMABLE

B

LIQUIDS

ELECTRICAL

C

EQUIPMENT

COMBUSTIBLE

D

METALS

When a fire starts, think first of your safety and the safety of others. Alert the fire department. Try to put out the fire only if you have been trained to use extinguishers, and the fire is small and tame enough to be extinguished by a hand-held extinguisher.

When the fire is out of control, the combustible material is unknown, or you have not been trained in the proper use of extinguishers, leave the fire fighting to professionals with the proper equipment. In this case, sound the fire alarm, then call for emergency help from a safe place.

Fire Checklist

Try not to panic. Although fire is a panic situation, when one panics, dangerous mistakes can be made. The calm person who assesses the extent of the blaze, calls the fire department, and acts quickly to contain or extinguish the blaze, is the one acting responsibly.

If the fire can be contained or extinguished, **a properly trained person should use the right extinguisher on the blaze.** When using a typical extinguisher, follow the "PASS" method. Hold the extinguisher upright, and:

- Pull the pin; stand back eight to ten feet.
- Aim at the base of the fire.
- Squeeze the handle.
- Sweep at the base of the fire with the extinguishing agent.

If you aim high at the flames, you won't put out the fire. Remember, too, that most extinguishers have a very limited operation time, only 8-10 seconds, so you have to act fast and spray correctly at the base of the fire, not at smoke or flames.

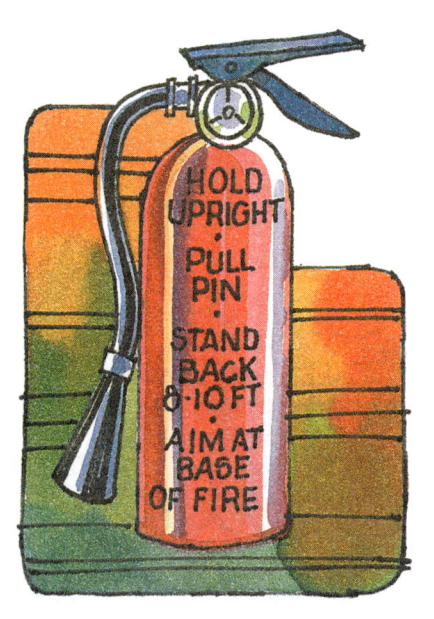

Time is of the essence in fire fighting. The smaller the fire, the easier it is to extinguish. Know the location of fire alarms and extinguishers. Know your nearest fire exit and proceed to it in an orderly fashion.

Be especially aware of smoke and noxious fumes. These fumes enter the lungs and leave a person unconscious. All fires consume oxygen to burn. Most victims of a fire suffocate from lack of oxygen and die. They are already unconscious or dead before the flames reach them.

Inside a building that is in flames, you should shut all doors within your reach. Get to your hands and knees and crawl to an exit. This is important because smoke and heat rise rapidly, and you will inhale less smoke near the floor. Outside, get away from the direction of the flames and smoke to avoid inhaling smoke and fumes.

Use a Shield

In any fire situation inside a building, anything you can use—any type of shield, blankets, or tarps—will help you get out of the building with less risk of injury. A wet cloth or handkerchief over your nose will help cut down the smoke intake.

Fire Prevention

On the job, it is part of your responsibility to help prevent fires. Extreme care is especially important when working with chemicals such as flammable solvents, gasoline, gases, and fuels.

The Chemical Fire

Many of the thousands of chemicals in use in the workplace are both highly toxic and highly volatile. Extreme caution must be used to prevent and fight fires resulting from chemical spills and accidents. Know the hazards of the chemical substances you use on the job and how to handle and store them properly to prevent dangerous chemical fires.

Chemical Hazards

Chemicals can cause serious injuries through physical (fire or explosion) or health (burns or poisons) hazards. Many chemicals have inherent properties that make them very hazardous. They might include:

- **Flammability** – These chemicals catch fire very easily; hazards include property damage, burns, and injuries.

- **Reactivity** – A reactive material is one that can undergo a chemical reaction under certain conditions; reactive substances can burn, explode, or release toxic vapor if exposed to other chemicals, air, or water.

- **Explosivity** – An explosive is a substance which undergoes a very rapid chemical change producing large amounts of gas and heat; explosions can also occur as a result of reactions between chemicals not ordinarily considered explosive.

As a result of these properties, chemicals can produce fires that start and spread quickly and may be difficult to fight or contain.

Fighting Chemical Fires

Unless you are a member of a fire fighting team, you will probably not be involved in battling a major chemical fire.

When fire extinguishers are used properly, they can and often do keep a small incident from becoming a major fire. However, you should be properly trained in their use and know their limitations. Remember that fire extinguishers are "first aid" appliances designed to answer immediate need. Early detection of a fire is essential if it is to be controlled with only an extinguisher. Call professional help immediately if the fire has spread out of control!

Flammable Liquid Handling and Storage

Flammable liquids give off ignitable vapors. Also, nearly all flammable liquid vapors are heavier than air and will accumulate in low areas with poor ventilation. When they accumulate sufficiently, they spread and can travel to an ignition source. These flames (or ignition sources) might be cigarettes, a hand tool that sparks, a cutting torch, or a motor.

The best way to stop fires in the workplace is to eliminate the conditions and practices that enable a fire to begin in the first place. This is why the handling and storage of flammable liquids is so crucial. Safety procedures and equipment for the safe handling of these liquids can be grouped into four segments. The basic safety principles apply to all of them. You may be involved in one or all aspects discussed in this section.

Storage

The typical plant stores flammable liquids in two ways: reserve storage in drums and operational storage in small quantities (for use at work stations.) For reserve storage safety, as soon as a drum is unloaded, the bung cap should be removed and a drum vent screwed in; this prevents pressure build-up if the drum is exposed to heat. Proper vents also incorporate emergency relief devices, which blow out under extreme pressure.

Drums should also be connected to a grounding system; this eliminates static electrical build-up when dispensing from the drum. If your plant does not have a drum storage room, drums should be stored in a safety cabinet; they are available in sizes to hold drums vertically or horizontally.

Transfer

Transfer of flammable liquids refers to their removal from storage to the places where they will be used. Liquids can be dispensed from drums by two methods: gravity flow from drums stored horizontally, and pumping from drums stored vertically.

For gravity flow safety, liquids should be dispensed into a safety can using a self-closing drum value. OSHA requires the use of approved safety cans for transfer purposes.

A drip can should be placed under the valve to catch spills and leaks. The drip can and receiving container must be bonded to the drum to draw off any static electrical charge.

The pump method is faster, empties the drum almost completely and saves space because drums are stored vertically. Drip cans are not required. Bond the receiving containers if the pump hoses are not self-bonding.

Mobile solvent tanks (liquid caddies) are used to distribute flammable liquids to work stations using large production line equipment. They are equipped with rubber wheels, a measuring pump, and a self-bonding hose.

Use

Use safety cans to hold and dispense flammable liquids as you work. There are many work station cans and tanks from which to choose. Liquids should be stored in safety cabinets at the work station. Keep containers closed when not in use.

Disposal

Disposing of waste flammable liquids requires as much caution in handling as do any of the other stages. Oily, solvent-soaked rags can easily start a fire. To prevent this, specially designed waste cans should always be used for temporary storage. These cans have spring-loaded lids and a raised bottom with vent holes to dispense heat. For removing flammable liquids from the work station for disposal, drain cans and liquid disposal cans offer the greatest degree of safety.

Spill Cleanup

It is the vapors, rather than the liquid itself, that burn. When the liquid is spilled, vapor release begins immediately, and continues until the liquid is removed. This requires that cleanup operations begin at once.

Specially developed absorbent materials have been developed for spill cleanup. These products are offered in pillows, pads, sheets, tubes and other shapes to fit all cleanup needs. Once the absorbent material is saturated, it should be placed in a large disposal drum and sealed with a drum cover. Another spill cleanup involves the use of specialized vacuum equipment.

Compressed and Liquefied Gases

The flash points of compressed flammable gases are extremely low and always below room temperature. Explosive mixtures are readily formed with air. Ignition of even a small leak may cause the materials to ignite.

To avoid fires resulting from ignition of compressed gases:

- Never roll or drag cylinders when gases are stored, transported, or used. Use a hand cart or truck specially designed for gas cylinders.

- Store all cylinders upright and secure them to walls or bench tops during storage or use.

- Compressed gases should be stored in dry, cool and well ventilated areas, protected from the weather, and away from flammable materials. The area should be posted for no smoking.

- Keep compressed gas cylinders which contain oxygen away from oil, grease, or liquid flammables.

- Separate fuel and oxidizing gas cylinders by at least 20 feet or a fire wall.

- When adequate ventilation can't be achieved, make sure safety equipment is at hand, including gas detectors, gas masks, self-contained breathing apparatus, and protective clothing.

- Be very careful about fittings or connections. Before any connections are made, inspect the cylinder carefully. Do not change, modify, repair, or tamper with pressure relief devices on cylinders.

- When more information, advice or help is needed, call the gas supplier; when in doubt about handling, contents or cylinder condition, seek an expert's advice.

Work at Working Safely

Any fire in the workplace has the potential to cause serious personal or property damage. When chemicals are involved, the possibilities for destruction are greatly multiplied. Prevention is the key to eliminating the hazards of any kind of fire where you work. Preparation is the key to controlling the consequences of a fire:

1. Keep work areas clean and clutter-free.
2. Know how to handle and store chemicals.
3. Know what you are expected to do in case of a fire emergency.
4. Call professional help immediately; don't let a fire get out of control (this applies to a fire wherever you are).
5. Know what chemicals you work with — you might have to advise fire fighters on the scene of a chemical fire concerning the type of hazardous substances involved.
6. Make sure you are familiar with your facility's emergency action plan for fires.

FIRST AID AND BLOODBORNE PATHOGENS: PROTECTING YOURSELF WHILE GIVING ASSISTANCE

The bloodborne pathogens standard was designed to provide a set of practices to follow when rendering first aid to help protect you against infections caused by germs carried in blood.

Where Are the Regulations?

Regulations governing exposure to bloodborne pathogens have been issued by the Occupational Safety and Health Administration (OSHA), specifically in 29 CFR 1910.1030. It is the employer's responsibility to develop an exposure control plan; provide training to those workers potentially exposed to bloodborne pathogens; implement engineering and work practice controls; enforce use of personal protective equipment; offer a hepatitis B vaccine, exposure evaluation, and follow-up; and use signs and labels to warn of potential hazards.

Key Definitions

Because of the technical nature of some of the words used when talking about bloodborne pathogens, some key definitions are spelled out here. Refer back to these definitions if you don't understand something later in this chapter.

Bloodborne Pathogens: Microorganisms present in human blood that can cause disease in humans. These include, but are not limited to, hepatitis B virus (HBV) and human immunodeficiency virus (HIV).

Exposure Incident: A specific eye, mouth, other mucous membrane, non-intact skin, or parenteral contact with blood or other potentially infectious material that results from doing one's job.

Occupational Exposure: A reasonably anticipated skin, eye, mucous membrane, or parenteral contact with blood or other potentially infectious material that may result from doing one's job.

Parenteral: A piercing of mucous membranes or the skin barrier by means of a needlestick, human bite, cut, and/or abrasion.

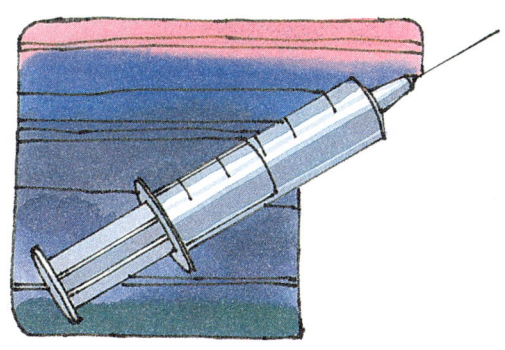

Universal Precautions: An infection control approach whereby all human blood and certain body fluids are treated as if they were known to be infectious for HIV, HBV, or other bloodborne pathogens.

Potentially infectious materials: Materials that may be present in a first aid emergency include blood, urine or other body fluids, and vomit, especially when you can see blood.

If you are a health professional, a designated first responder or first-aid provider in your company, or involved in maintenance or housekeeping work that could potentially expose you to bloodborne pathogens, you need to know about this standard.

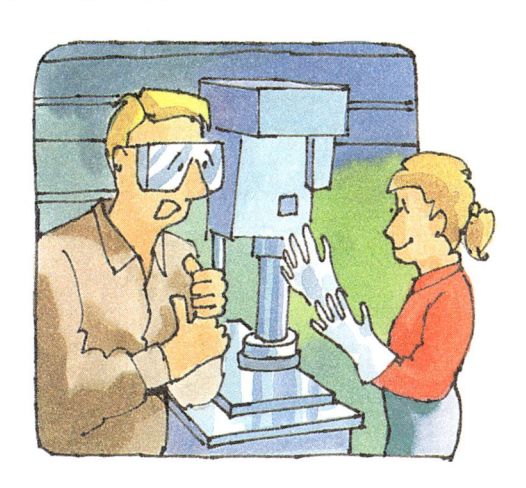

FIRST AID

Other types of workers covered by the standard include healthcare workers, police officers, firefighters, and employees of correctional facilities and funeral homes.

"Good Samaritan" acts performed by undesignated employees are not covered by the standard, but undesignated first-aid responders may want to know exposure controls anyway, to protect themselves if they voluntarily respond in the event of an emergency.

Required Elements of the Standard

When workers are exposed to blood or other potentially infectious material on the job, the standard requires that employers develop an exposure control plan.

Employers must also provide training on the following subjects to affected workers:

- Bloodborne diseases and how they are spread.
- The exposure control plan.
- Engineering and work practice controls.
- Personal protective equipment.
- Hepatitis B vaccine, exposure evaluation, and follow-up.
- How to respond to emergencies involving blood.
- Signs and labels used to warn of potential hazards.

FIRST AID

Exposure Control Plan

While exposure control plans will vary from workplace to workplace, they have some common elements:

- Identification of job classifications and, in some cases, tasks where there is exposure to blood and other potentially infectious materials.

- A schedule of how and when the provisions of the standard will be implemented, including schedules and methods for communication of hazards to employees, hepatitis B vaccination and post-exposure evaluation and follow-up, recordkeeping, and implementation of:

    - engineering and work practice controls.
    - personal protective equipment.
    - housekeeping.

- Procedures for evaluating the circumstances of an exposure incident.

If you are an "affected employee," it would be a good idea for you to ask your supervisor for a copy of the exposure control plan, read it, and be sure you understand it.

Engineering Controls and Work Practices

The employer must also institute engineering controls and work practices that will minimize the possibility of exposure. Such things as handwashing, prevention of needlesticks, and minimization of the splashing or spraying of blood fall under this category. Engineering controls eliminate hazards at their source. This includes the use of autoclaves and containers for used sharps. Engineering controls must be checked and maintained on a regular schedule to keep them in good working order. Employers must provide accessible facilities for handwashing.

Wash hands immediately after removing gloves or other protective equipment, and after any hand contact with blood or potentially infectious fluids. If a sink isn't available for handwashing, antiseptic cleansers must be provided. In this case, wash with soap and water as soon as possible.

Do not bend, shear, break, remove or recap any used needle or sharp. Dispose of used sharps in the proper containers. These containers must be puncture resistant, be properly labeled, and have leakproof sides and bottoms.

Eating, drinking, applying cosmetics or lip balm, and handling contact lenses are prohibited in areas where there is a potential for exposure. Food or drink cannot be stored in refrigerators, freezers, shelves, cabinets or on countertops where blood is stored or where blood or other potentially infectious materials may be present. Procedures involving blood or potentially infectious material must be performed in a manner that reduces spraying or splashing to a minimum. Blood or fluid specimens must be placed in a container that does not leak during handling, storage, or shipping.

Personal Protective Equipment

Employers must provide, and employees must use, personal protective equipment (PPE) when the possibility exists for exposure to blood or body fluids. This equipment must not allow blood or potentially infectious matter to pass through it to the employee's clothes, skin, eyes, or mouth.

Personal protective equipment must be accessible and available in appropriate sizes. (Hypoallergenic or powderless gloves must be available to those allergic to regular gloves.) PPE also must be kept clean and in good repair.

Single use gloves must be replaced as soon as possible after they are contaminated or if they become torn or punctured.

These gloves should never be reused. Various other types of PPE include plastic visors, half-face masks, full body gown, goggles, etc.

Whatever it takes to stop exposure to blood or other potentially infectious materials is the level of PPE you should be wearing when you provide first aid.

Housekeeping Techniques

Housekeeping staff may be occupationally exposed to potentially infectious material if they clean up after some first aid incidents. Therefore, housekeeping staff need to follow the written cleaning schedule that outlines the method of decontamination to be used and describes the proper disposal of used sharps.

Equipment and work areas must be cleaned and decontaminated as soon as feasible after contact with any blood or potentially infectious fluids.

Protective coverings must be removed and replaced when overtly contaminated or at the end of each shift if there is a possibility of contamination during the shift.

Contaminated laundry should be handled as little as possible. Laundry must be bagged where it was contaminated. Wet laundry must be placed in leak-proof bags.

All employees who handle contaminated laundry must wear gloves.

Hepatitis B Vaccine

The greatest bloodborne risk is infection by the hepatitis B virus. Because of this risk, your employer must make hepatitis B vaccine available to you when you are exposed to blood or other potentially infectious materials on the job.

Pre-screening cannot be done as a condition of receiving the vaccine. If you refuse to be vaccinated, you must sign a declination form. If you change your mind later on, your employer must still provide the vaccine.

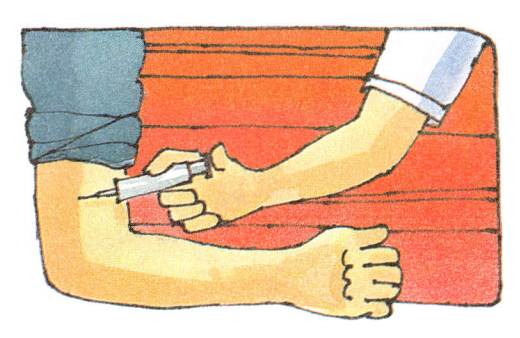

Exposure Incident

An exposure incident is a specific eye, mouth, other mucous membrane, non-intact skin, or parenteral contact with blood or other potentially infectious material that results from doing one's job or providing first aid as a first responder. When an exposure incident is reported, the employer will arrange for an immediate and confidential medical evaluation. The medical evaluation must:

- Document how the exposure occurred.
- Identify and test the source individual if feasible.
- Test the exposed employee's blood, if consent is obtained.
- Provide counseling.
- Evaluate, treat, and follow up on any reported illness.

Your employer must provide the medical professional doing the exposure assessment with all relevant data needed to complete the employee's evaluation.

Communication of Hazards

All warning labels must bear the biohazard symbol, be printed in fluorescent orange or orange-red, and have lettering of a contrasting color.

Red bags or containers may be used as a substitute for labels. Labels must be placed as close to the container as possible on all packages of regulated waste, refrigerators/freezers containing blood or other potentially infectious material, and other containers used for shipping or storing blood and body fluids.

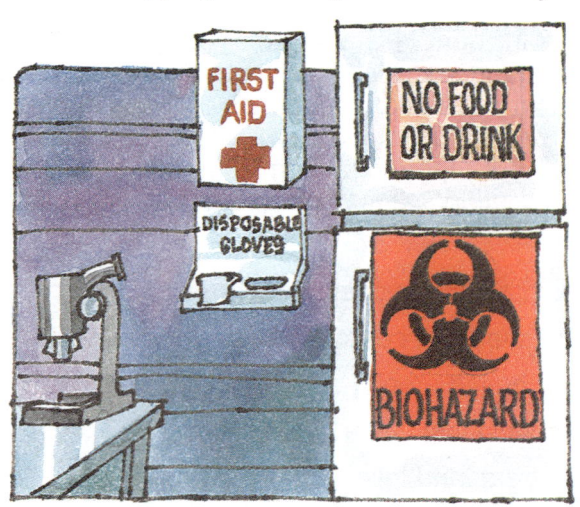

Blood that has been tested and found free of HIV or HBV and released for clinical use and decontaminated regulated waste do not require labels.

Recordkeeping

Records must be maintained on all employees with occupational exposure for the period of their employment plus 30 years.

Each record, which must be available to the employee, should include:

- Name and social security number.
- Hepatitis B vaccination status.
- Results of all exams, testing and follow-up procedures.
- Copy of healthcare professional's opinion.
- Copy of information provided to healthcare professional.

Note: These records are confidential and can be released only with the employee's written consent or if required by law.

In addition, training session records must be kept for three years. Training records must include:

- A summary of program contents.
- Dates training occurred.
- Trainer's name and qualifications.
- Names and job titles of all participants.

Rendering first aid is a wonderful life-giving thing to do. But if you don't protect yourself in the process, you risk exposing yourself to harmful and sometimes deadly bloodborne pathogens. To protect yourself, follow the bloodborne pathogens program your employer has set up.

Work at Working Safely

Remember the key elements of a bloodborne pathogens program. If you're a first-aid provider, maintenance or janitorial person responsible for cleaning up potentially infectious materials, or any other potentially exposed employee, then you need to know about the following:

1. The written exposure control plan.
2. The training to be provided to you.
3. Engineering controls and work practices to minimize chance of exposure.
4. Personal protective equipment to provide barriers to exposure.
5. Housekeeping techniques to protect you.
6. The hepatitis B vaccine you are entitled to.
7. The use of labeling and red bags to indicate contaminated waste.
8. Steps to be taken in the event of an exposure incident.

FORKLIFT SAFETY: TIPS FOR DESIGNATED DRIVERS

Dangerous forklift drivers are found across the country, in almost every plant and warehouse. Forklift accidents have increased dramatically in recent years. Most occur when the driver hits a co-worker because of an error in judgment. Few accidents result from overturning the forklift, but when these occur, they often result in serious injuries and even fatalities.

Take Forklift Driving Seriously

Driving a forklift is a serious responsibility. Consider that the average car weighs between 2,500 and 3,500 pounds. A 6,000 pound capacity forklift weighs two or three times as much. With a capacity load, you are handling a mass as high as 16,000 pounds. A lift truck with eight tons of mass that travels at a high rate of speed can cause tremendous damage should something be hit or run over.

Where Are the Regulations?

The Occupational Safety and Health Administration (OSHA) has issued regulations specifying general requirements for powered industrial trucks (OSHA's term for forklifts, platform lift trucks, motorized hand trucks, among others) safety and training. These regulations can be found in 29 CFR 1910.178.

According to OSHA's 1998 forklift operator training final rule, operators must successfully complete training and an evaluation of their skills.

The training covers safe operation of the type of truck that is being used, the hazards in the workplace, and OSHA's regulation.

Drivers will have refresher training and evaluations as needed. An evaluation of each operator's performance must be done at least every three years.

Forklifts: Basic Loading/Unloading Tools

Forklifts are wonderful tools for moving and stocking material. These trucks are powered by various means: gasoline or diesel fuel, propane gas, or electric power from batteries.

Some are made as stackers moving on rails, and some are shovel loaders. Lift trucks come with a number of specialized options, such as clamps, pole carriers, buckets, and swing arms. There are related lifters, such as pallet trucks, towing trucks, low lifters, and port-trucks. All require special training and knowledge of safe operation.

With the variety and styles of powered lift trucks available, it is everyone's business to be aware of the hazards and safety requirements for safe operation. Safety awareness is important for the driver and for others working in a warehouse, because anyone can become a victim of an unsafe driving act.

Forklifts Are Unique

Driving a forklift is a unique experience. Forklifts weigh much more than cars and are steered from the rear axle rather than the front as in a car. Rear wheel drive allows a forklift to turn in a tight radius, but also means that a forklift will swing considerably more than the rear of a car when turning.

Operating a forklift is fundamentally different from driving cars or other trucks. Be aware that a forklift:

- Is usually steered by the rear wheels.
- Steers more easily loaded than empty (because the load is balanced by counterweights).
- Is driven in reverse as often as forward.
- Is often steered with one hand.
- Has a center of gravity that is towards the rear and shifts toward the front as the forks are raised.

Cars use a four-point suspension system while forklifts use a three-point suspension. This system permits the center of gravity to shift in a forklift but makes it more likely to tip over.

Center of Gravity

The center of gravity for a forklift moves depending on the load and how it is positioned. The center of gravity will move when accelerating, braking, and turning. Therefore, it is very important to avoid quick accelerating or braking or turning a corner too fast.

Another factor that will affect the center of gravity is the load itself. Position the load close to the mast and tilted back. Tilting the mast back when traveling with a load creates better vehicle/load balance.

Never travel with the load elevated any more than is necessary to clear any bumps or curbs. On an incline of more than ten percent, drive with the load upgrade, forward up a ramp and reverse down a ramp.

Remember that these units are made to travel with loads. An unloaded forklift has the potential to tip because of the extreme weight of the counterbalance. Safe driving is just as important with an unloaded forklift as it is with one that is loaded.

Nameplates

Each manufacturer provides an identification plate, or nameplate, on every forklift they build. This plate gives you valuable information about the forklift's design and capacity. It tells you how much the unit weighs (important to know before you drive into a trailer or elevator) and how much it can carry.

Forklift Accidents Can Be Prevented

Drivers need to be trained before they operate a lift truck. Not only is it important to know how to professionally operate a forklift, it's vital to know all safety rules of operation.

Become familiar with these rules before you think about driving a forklift.

1. Operating a forklift takes skill, mechanical knowledge, compliance with safety rules, and defensive driving under unique conditions.
2. **Only the assigned driver or drivers, should operate a forklift.** An assigned operator is responsible for the cleanliness, maintenance, and security of the vehicle.
3. If at any time a forklift is found to be in need of repair, defective, or in any way unsafe, **the truck must be taken out of service until it is safe to operate.**
4. **Only a trained and authorized operator should drive a forklift.** A qualified operator is one who has been fully trained and tested, knows the general vehicle design, and who has learned safety inspections and safe driving rules.

Operating Your Forklift Safely

Start by getting into and out of your forklift properly. Use a three-point stance with two hands and one foot in contact with the floor or unit at all times. Never jump on or off the forklift.

Become familiar with all controls (both location and operation). Controls may vary from unit to unit. Be sure you understand every control on your forklift before you start it.

If the forklift has a seat belt, wear it! This is an important safety feature that may protect you during an accident. A seat belt will help to hold you in the frame of the safety cage should the vehicle tip. If you attempt to jump off the forklift, you are likely to be trapped under it or the load it's carrying.

Wear protective equipment when required, such as safety glasses and ear protection.

The following "rules of the road" list general guidelines for safe forklift operations.

- Always keep arms and legs inside the vehicle.

FORKLIFT

- Face direction of travel, keep your mind on what you are doing, and never travel forward with the load blocking your view.
- Keep three vehicle lengths away from other vehicles – it's a space cushion around the vehicle.
- No horse play is allowed. It's basic common sense.
- Be aware of overhead clearances, such as pipes, sprinklers, door beams, and know the load limits of elevators.

When picking up a load:

- Make sure the load does not exceed the capacity of your forklift.
- Make sure forks are positioned properly.
- Make sure the load is balanced and secure.
- Check for overhead obstructions.
- Raise the forks to the proper height.
- Bring the forks all the way into a pallet, and tilt the mast back to stabilize the load before moving.
- Back out, stop completely, then lower the load.

Traveling with a load:

- Pedestrians always have the right of way.
- Never allow anyone to ride on your forklift.
- Always watch where you are going.
- Keep the forks low.
- When moving, always have the unloaded forks as low as possible, but high enough to clear bumps and curbs. Never travel with a load raised high.
- Know the position of your forks at all times.
- Keep the load tilted back slightly.
- Obey speed limits. Remember that a forklift is not a street rod, but is a slow moving vehicle, designed that way for safety.
- Slow down at all intersections, and always sound the horn at blind ones.

- Always drive up and back down ramps and inclines.
- Avoid sudden braking.
- Lift or lower the load only when completely stopped, never when traveling.
- Keep to the right, the same as highway driving with a car.
- Be alert for oil and grease spots, which could reduce traction.
- Make sure the load is balanced and secure on the forks.
- Cross railroad tracks at an angle, one wheel at a time.
- Be careful of changing light conditions, such as coming in from bright daylight into dimly lit areas, and vice-versa.
- Beware of edges on loading docks.

Whether you are moving a load or just driving the forklift, your visibility is restricted. Be very careful to watch your blind spots. Accidents involving forklifts that hit pedestrians and objects are far too common.

When placing a load:

- Stop the forklift completely before raising the load.

FORKLIFT

- Move slowly with the load raised.
- Never walk or stand under a raised load.
- Tilt the load forward only when over a stack or rack.
- Be certain the forks clear the pallet before turning or changing height after you've set down the load.
- Always stack the load square and straight.
- Before backing, check behind and on both sides for pedestrians or other traffic.

When you leave a forklift unattended, completely lower the load engaging means, neutralize controls, set the brakes to prevent movement, and shut off the power. **NOTE:** A powered industrial truck is unattended when the operator is 25 feet or more away from the vehicle which remains in his or her view, or whenever the operator leaves the vehicle and it is not in view.

If you get off the truck and stay within view and within 25 feet of it, you must still lower the forks, put the controls in neutral, and set the brakes; but you can leave the power on.

Fueling the Unit

Refueling raises still more safety issues. Be sure to carefully follow the procedures set up for your lift trucks whether they are powered by propane, gasoline or diesel fuel, or batteries.

Vehicle Maintenance Is a Continual Job

Maintaining your forklift properly is just as important as driving safely. A regular maintenance schedule should be set up for forklifts, and you should always run down a safety checklist at the start of your shift. This check includes a visual inspection, as well as checking fluid levels, hydraulics, wheels and tires, brakes, and any potential mechanical problems on the vehicle. Any forklift not in safe operating condition must be removed from service.

Brakes

The most important of your inspections, **brakes are the single most common cause of lift truck accidents due to mechanical failure.** Push the brake pedal in. It should have free travel before meeting resistance. Then, depress the pedal again and hold it for ten seconds. The pedal must hold solid and not be spongy or drift under pressure. Make sure the parking brake and the seat brake on electric trucks are working properly. If the brakes are not working properly, the vehicle should not be driven and the problem reported.

Steering

Steering is a vital maintenance concern. **With the engine running, check if the steering wheel turns correctly both ways to its stops.** The wheel should not feel loose and the pump should not squeal before reaching the stops.

Adhere to this checklist before operating a vehicle.

- Check the fork pins and stops to make sure that they are in place.
- Check all cowling and body parts.
- Check the fuel level, crankcase oil level, radiator water level. Check the engine air cleaner, the fan belt, the hydraulic fluid level, and the battery water level.

- Check the hour meter and record it. This is important for maintenance scheduling.
- With the engine running, check operation of the hour meter, headlights, taillights, and warning lights.
- Check the oil pressure gauge, the water temperature, ammeter, and sound the horn. Note if the clutch is working properly. Check the hydraulic controls and any other controls on the lift system.
- Check the wheels and tires for excessive wear.
- Look for any broken or loosened parts.

These checks are not excessive. For safety's sake, and for your own well being, you need to know if your vehicle is safe to operate. Anything not up to par must be reported to your foreman at once.

Getting Your Attention

Human error is the primary reason for most lift truck accidents. The greatest cause of accidents among new drivers is forgetting to watch for overhead obstructions when lifting a load.

FORKLIFT

Other causes that rank high on the list are driving on the wrong side of the aisle and daydreaming.

Remember, when operating a forklift, all of your attention must be focused on what you are doing. When your mind is a million miles away or when you are not paying attention, you are at the greatest risk of having an accident. You can get the job done properly by thinking and driving defensively. Know what you are doing when driving a forklift truck.

Work at Working Safely

1. Never park in front of fire equipment, doors, exits, or high traffic areas.
2. Do not pass another vehicle in narrow aisles.
3. Never smoke in fueling areas.
4. If you cannot see past a load in front, travel backwards, carefully.
5. Know the load capacity and limits of your vehicle. Also, stay within any elevator or floor load limits.
6. Never attempt to lift a load beyond the load limits of your forklift.

FORKLIFT

7. Do only maintenance or repair work that you are authorized to do. Leave the rest to maintenance personnel.
8. When leaving your vehicle, lower the forks, put the controls in neutral, set the brakes, block the wheels if on an incline, shut the power off, and remove the ignition key or connector plug.

When driving a forklift, your responsibilities are great. By using common sense, you can head off potential hazards and accidents. You can do that if you know what to watch for, drive defensively, and make sure that you are being the professional that you should be. Think, act, and be professional in your forklift driving. Be a skilled and safe driver, able to handle the job efficiently and safely.

HAND TOOLS: THE RIGHT TOOL FOR THE JOB

Some jobs would be impossible without using hand tools. Everyone is familiar with common, everyday hand tools, but don't take them for granted. Hand tool mishaps can cause serious injuries. If you know how to use the tools and take care of them, you'll have a better chance of avoiding an injury.

Avoid Tool Injuries

There are many types of injuries that can be caused by tool use. Serious eye injuries can result if materials shatter while using hammers or mallets. Filing or chiseling creates chips that can get into your eyes. While you're looking up to use tools above your head, dust and debris can fall into your eyes. A screwdriver could slip and cause a puncture wound. If a knife slips, you could cut a tendon, artery, or nerve. A misplaced hammer blow can break a bone in your hand. Tool use presents plenty of other opportunities for minor scrapes, cuts, or bruises.

Injuries can be prevented by keeping tools in good condition, using the right tool for the job, using the tool properly, setting up the work so you aren't straining yourself, and wearing personal protective equipment. Always wear safety glasses when using hammers, chisels, punches, wire cutters, saws, files, crowbars, bolt cutters, or any tool that could create chips or pieces. Wear cut resistant gloves when handling knives or other sharp edges. Arrange the work and use tools so that the tool will move away from your hands and body if it slips. Make sure that the material you are working on is held securely—use clamps or a vise if you need to. And, stand where you have firm footing and good balance while you use tools.

Select the Tool You Need

Use durable tools made from good quality materials. Metal tools should have working points that resist bending, cracking, chipping, or excessive wear from normal use. Handles should be made of durable material that does not crack or splinter easily if the tool is dropped or hit. You may need a handle with a cushioned grip to help absorb impacts or squeezing pressure. Consider the temperature of the work environment and select tools with handles that will insulate you from temperature extremes. Selecting tools with comfortable grips eases strain and gives you better control over the tool.

Use tools that will contribute to your safety during hazardous jobs. Pay extra attention to any tools that you will be using around electrical parts. Make sure that the handle is electrically insulated and rated to handle the voltage. If you need to use tools to work in areas where flammable liquids are stored or used, the tools must be made from non-sparking alloys in order to prevent sparks that can ignite flammable vapors.

Use the Right Tool the Right Way

Look at your hand tools. Their shape and design shows you how they are intended to be used.

Knives: Using knives as pries, screwdrivers, can openers, awls, or punches can easily damage the blade. A sharp blade needs less pressure to cut and has less of a chance of getting hung up and slipping. Always move the blade away from you as you cut.

Screwdrivers: Using screwdrivers as pries, can openers, punches, chisels, wedges, etc. can cause chipped, rounded, bent, dull tips; bent shafts; and split or broken handles. If the screwdriver tip doesn't fit the screw, you'll apply more force and the screwdriver can easily slip. Redress the tips of flat head screwdrivers to keep them sharp and square edged. Screwdrivers with shorter shafts give you better control. Screwdrivers with thicker handles apply more torque, with less effort on your part.

Hammers and Mallets: Nail hammers are designed to drive nails. Ball pein hammers are designed for striking cold chisels and metal punches. Mallets have a striking head of plastic, wood, or rawhide and are designed for striking wood chisels, punches, or dies. Sledgehammers are for striking concrete or stone. You can damage a hammer by trying to use it for the wrong purpose. Don't use a hammer with a mushroomed striking surface or a loose handle. You can damage other tools by trying to force them by hitting them with a hammer.

Pliers: Don't substitute a pliers for a wrench. The face of the pliers is not designed to grip a fastener, and the pliers can easily slip off of the nut or bolt. Pliers are designed for gripping so you can more easily bend or pull material. They'll provide a strong grip if you protect them from getting bent out of shape and keep the gripping surface from being damaged.

Cutters: Use cutters or snips to remove banding wire or strapping. Trying to use a pry bar to snap open banding can cause injuries. Keep the cutting edges sharp and protect them from getting nicked or gouged.

Wrenches: Use adjustable open-ended wrenches for light-duty jobs when the proper sized wrench isn't available. Position yourself so you will be pulling the wrench towards you, with the open end facing you—this lessens the chance of the wrench slipping off of the fastener when you apply force. Select an open-ended wrench to fit the fastener for medium-duty jobs. With the snug fit, these wrenches can apply more force than an adjustable open-ended wrench. Again, pull the wrench with the open end facing you to avoid slippage. Box and socket wrenches should be used when a heavy pull is required. Because they completely encircle the fastener, they apply even pressure with a minimal chance of slipping. Some box wrenches are designed for heavy-duty use, and they do have a striking surface. But, in general, don't try to increase the torque by hitting the wrench with a hammer or by adding a cheater bar to the wrench's handle—this can break or damage the wrench. If the fastener is too tight, use some penetrating oil to lubricate it.

Wood Saws: For cutting wood, use a cross-cut saw to cut across the grain, and use a ripping saw to cut with the grain. Select a saw with coarse teeth for sawing green wood, thick lumber, or for making coarse cuts. Fine-toothed saws can be used to make fine cuts in dry wood. After use, wipe the saw with a lightly oiled rag to keep the teeth clean. Protect the saw from getting bent or damaged in storage.

Metalworking Hand Tools: Hack saws should have the blade installed with the teeth facing forward, and apply pressure on the forward stroke. Use a light pressure to avoid twisting and breaking the blade. Metal files need to be kept clean and protected from damage. Hitting the file against a hard object to clean it can damage the file—use a file card for cleaning.

Maintain Your Tools

Tools will last longer when you take care of them:

- Inspect tools before you use them and before you put them away.

- Maintain and repair tools before it's too late. Sharpen cutting edges regularly. Avoid waiting until they are completely dull before trying to sharpen them. Replace loose handles on hammers or mallets.

- Keep tools clean. Grease and dirt can hide damage, and prevent you from getting a good grip when you use the tool.

- Discard damaged tools. Striking tools with mushroomed surfaces, screwdrivers with rounded edges or bent shafts, or bent pliers are examples of tools that can cause more harm than good.

HAND TOOLS

Work at Working Safely

- Take out only the tools that you will need for the job. Piles of extra tools can get in the way or get lost.
- Carry your tools safely. Use a tool box or a tool chest to move tools around. If you need to carry tools, especially on a ladder, wear a tool belt. If you are working on a platform or a ladder, keeping the tools in your toolbelt helps keep them from being dropped onto unsuspecting victims below.
- Protect yourself and your tools. Keep knives in sheaths. Putting tools away after use keeps them from getting damaged or disappearing.
- Always wear appropriate personal protective equipment.

HAZARD COMMUNICATION: THE RIGHT TO KNOW LAW

If you come into contact with hazardous chemicals in your workplace each day, you are definitely not alone. One out of every four workers contacts hazardous chemicals on the job. In many cases, the chemicals you deal with may be no more dangerous than those you use at home. But in the workplace, exposure is likely to be greater, concentrations higher, and exposure time longer. Thus, potential danger is greater on the job.

Where Are the Regulations?

The Occupational Safety and Health Administration (OSHA) has issued a regulation to help control chemical exposure on the job. The regulation is called the hazard communication standard, but is more commonly called "hazcom" or the "Right to Know Law." It can be found at 29 CFR 1910.1200.

The standard says you have a right to know what chemicals you are working with or around. It is intended to make your workplace safer. So it's important that you are aware of the standard and the rights it grants you.

The hazcom standard requires that all chemicals in your workplace be fully evaluated for possible physical or heath hazards. And, it mandates that all information relating to these hazards be made available to you.

Who and What Does the Standard Cover?

The hazard communication standard really involves just about anyone who comes into contact with hazardous chemicals. Everyone needs to know what hazardous chemicals they work with and how to protect themselves. You and your co-workers will learn about the chemicals you work with and how to take precautions against any potentially negative effects associated with them.

Both training and written materials will inform you about chemicals you work with. Ask your supervisor about any questions you might have when looking at material safety data sheets (MSDSs) or the written program.

The areas specifically covered in the standard include:

- Determining the Hazards of Chemicals
- Material Safety Data Sheets (MSDSs)
- Labels and Labeling

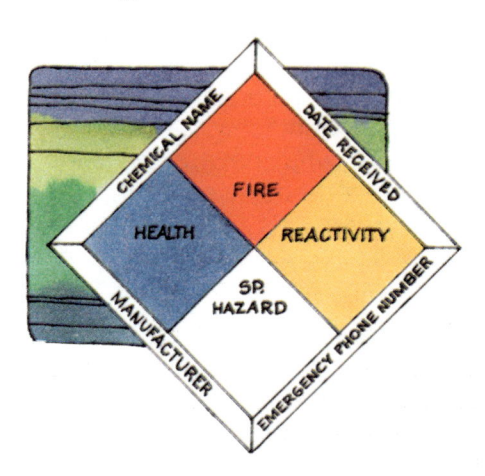

- A Written Hazard Communication Program
- Employee Information and Training
- Trade Secrets

The hazard communication standard is intended to cover all employees who may be exposed to hazardous chemicals under normal working conditions or where chemical emergencies could occur. As mentioned previously, the standard applies to those chemicals which pose either a physical or health hazard.

What Are Physical and Health Hazards?

Physical hazards are exhibited by certain chemicals due to their physical properties – flammability, reactivity, etc. These chemicals fall into the following classes:

- Flammable liquids or solids

- Combustible liquids
- Compressed gases
- Explosives
- Organic peroxide
- Oxidizers
- Pyrophoric materials (may ignite spontaneously in air at temperatures of 130°F or below)

- Unstable materials
- Water-reactive materials

A *health hazard* is a chemical that may cause acute or chronic health effects after exposure. It can be an obvious effect, such as immediate death following inhalation of cyanide. But a health hazard may not necessarily cause immediate, obvious harm or make you sick right away. In fact, you may not see, feel, or smell the danger.

An acute health effect usually occurs rapidly, following a brief exposure. A chronic health effect is long, continuous and follows repeated long-term exposure.

What Kinds of Chemicals Cause Health Hazards?

Some examples of chemicals which exhibit health hazards are:

Type of chemical	Example
Carcinogens (cause cancer)	formaldehyde or benzene
Toxic Agents	lawn and garden insecticides, arsenic compounds
Reproductive Toxins	thalidomide or nitrous oxide
Irritants	bleaches or ammonia
Corrosives	battery acid, caustic sodas
Sensitizers	creosote or epoxy resins
Organ-Specific Agents (act on specific organs/parts of body)	sulfuric acid (affects skin), or asbestos (affects lungs)

The hazard communication standard doesn't apply to hazardous waste/substances regulated by the Environmental Protection Agency, biological hazards, tobacco products, many wood or wood products, or food, cosmetics, or certain drugs.

The Material Safety Data Sheet

A material safety data sheet (MSDS) is a fact sheet for a chemical which poses a physical or health hazard in the workplace. MSDSs must be in English and contain certain information:

- Identity of the chemical (as used on the label).
- Physical and chemical characteristics (e.g., vapor pressure, flash point).
- Physical hazards.
- Health hazards.
- Primary routes of entry.
- PELs, TLVs, or other exposure limit used or recommended by MSDS preparer.
- Whether it is a carcinogen.
- Precautions for safe handling and use.
- Control measures (e.g., engineering controls, work practices).
- Emergency and first-aid procedures.
- Date of preparation of latest revision.
- Name, address, and telephone number of manufacturer, importer, or other responsible party.

Your organization must have an MSDS for each hazardous chemical it uses. Copies must be kept where you can use them during your workshift. When employees must travel between workplaces during the day, MSDSs may be kept at a central location.

If relevant information in one of the categories was unavailable at the time of preparation, the MSDS must indicate that no information was found. Blank spaces are not permitted. If you find a blank space on an MSDS, contact your supervisor.

Labels and Labeling Requirements

Containers of hazardous chemicals must be labeled in English. Information may also be presented in other languages for non-English speaking employees, but English is required. It is required that labels contain the following information:

- Identity of the hazardous chemical.
- Appropriate hazard warnings.
- Name and address of the chemical manufacturer, importer, or other responsible party.

On individual stationary containers you may use signs, placards, batch tickets, or printed operating procedures in place of labels.

Some employers opt to standardize their labels for hazardous chemicals. This requires relabeling each container of a hazardous substance as it enters the plant. Some organizations choose to do this because their employees and any emergency responders then have a uniform labeling format to use with easily recognizable hazard warnings.

Many systems for labeling hazardous chemicals have been developed such as the National Fire Protection Association's system or the Hazardous Material Information System. OSHA does not require relabeling (unless a label falls off or becomes unreadable), nor does it require any particular system.

Written Hazard Communication Programs

Your employer is required by the hazard communication standard to have developed and implemented a written hazard communication program. This program details how the employer will meet the standard's requirements for labels, MSDSs, and employee information and training.

Your organization's written program needs to include:

- A list of the hazardous chemicals known to be present in your workplace.
- How the MSDS requirements are being met.
- What type of labeling system, if any, is used.
- Detailed information on training compliance.
- Methods your employer will use to inform you of the hazards of non-routine tasks and such things as unlabeled piping.
- Methods your organization will use to inform other employers of workers on your site, such as service representatives, repairmen, and subcontractors.

You Must Be Trained

You must be trained at the time of your initial employment or assignment, as well as whenever a new hazard is introduced into your workplace.

According to the hazard communication standard, you are to be informed of the requirements of the standard. You are to be informed of any operations in the work area where hazardous chemicals are present.

You also need to be informed of the location and availability of your facility's written hazard communication program. Even more important, the location and availability of the MSDS file should be stated.

Your training must contain all of the following elements:

- **Methods or observations** used to detect the presence or release of hazardous chemicals in your work area.
- **Physical/health hazards** of chemicals in your work area.
- **Measures you can take to protect yourself** from hazards, including work practices and personal protective equipment.
- **Details of your employer's hazard communication program,** including complete information on labels and MSDSs.

Work at Working Safely

Training is the key to your success and safety as an employee dealing with hazardous chemicals in the workplace. Take it seriously. Get as much as you can from it. Learn about MSDSs, labeling, your employer's written program, measures to protect yourself, and what hazardous chemicals you work with. Your good health may depend on how much you learn from your organization's training program.

LIFTING TECHNIQUES: AVOIDING BACK INJURIES

Sprains and strains are the most comon causes of lower back pain. Your back can be injured by improper lifting of moderate to heavy objects, falling, auto accidents, and sports activities. But of these, lifting improperly is the largest single cause of back pain and injury. Luckily, you can do something about preventing back pain by knowing and using proper lifting techniques.

Along with the common cold, problems with the lower back are a frequent cause of lost work time and worker's compensation. The Bureau of Labor Statistics (1997) reported over 311,900 injuries resulting in days away from work (representing 16.5 percent of all occupational injuries or illnesses) from overexertion during 1996. Not only does industry lose, but you lose if you're laid up for weeks, unable to stay active.

LIFT

Although our backs hold up well, our lifestyles and activities can lead to back pain. Here are some things that can go wrong.

- **Strains and sprains** can result from injury to muscles and ligaments that support the back. A torn ligament will result in severe back pain.
- **Ruptured or slipped disk** is not uncommon and occurs when the disk (vertebral cushion) presses on a nerve.
- **Chronic tension or stress** can result in muscle spasms and aggravate persistent and painful backache.
- **Other conditions** such as pain "referred to the back" from other organs, such as the kidneys and prostate, can result in nagging back pain.

Why Back Pain Happens

Using improper lifting techniques can lead to back injuries, but other factors can contribute to this age-old problem.

Poor Posture

Whether you're standing, sitting, or reclining, posture affects the amount of strain put on your back. The wrong posture increases strain on the back muscles and may bend the spine into positions that will cause trouble. When standing correctly, the spine has a natural "S" curve. The shoulders are back and the "S" curve is directly over the pelvis.

Good sitting posture should put your knees slightly higher than your hips. Your hips should be to the rear of the chair with your lower back not overly arched. Also, your shoulders and upper back are not rounded. Reclining posture is important, too. Sleep on your side with knees bent or sleep on your back. Sleeping on your stomach, especially on a sagging mattress with your head on a thick pillow, puts too much strain on the spine. Result: morning backache.

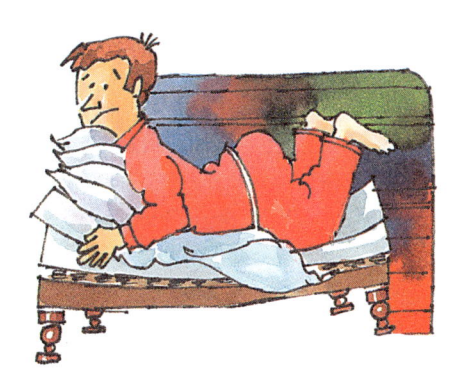

Poor Physical Condition

Your physical condition can lead to back pain. If you are overweight, and especially if you have developed a pot belly, extra strain on your spine results. An estimate is that every extra pound up front puts 10 pounds of strain on your back.

When you are out of shape, the chances for chronic back pain are greater. Infrequent exercise is a major factor, too. A sudden strain on generally unused back muscles leads to trouble, particularly when there is a sudden twisting or turning of the back. Proper diet and exercise is the sensible way to help avoid back problems.

LIFT

Stress is another factor that may lead to back pain. Tied in with your general physical condition, stress created from work or play can cause muscle spasms that affect the spinal nerve network. Although stress is part of everyone's life, and a certain amount of stress is normal, excessive stress causes backache. The solution is a balanced life style with time to relax.

Repetitive Trauma

People often think back injuries result from lifting heavy or awkward objects. Many back injuries, however, do not come from a single lift, but occur from relatively minor strains over time.

Back injuries, as with other cumulative trauma disorders (CTD), may arise from repeated injuries. As the worker repeats a particular irritating movement, the minor injuries begin to accumulate and weaken affected muscles or ligaments. Eventually a more serious injury may occur.

Thus, a specific weight lifted may actually have little to do with any single injury. Remember to use mechanical aids when appropriate along with good lifting techniques, whenever you do any lifting. You can lift safely when you lift with caution.

Basics of Good Lifting

Today, most heavy objects are lifted by forklifts, hoists, dollies, and other types of equipment. However, sometimes it is necessary to load or unload moderate to heavy objects by hand.

When that is the case, knowing the proper ways to lift can save you a great deal of pain and misery from a sprained back.

1. **Size up the load before trying to lift it.** Test the weight by lifting at one of the corners. If the load is too heavy or of an awkward shape, the best thing to do is:
   - Get help from a fellow worker.
   - Use a mechanical lifting device.
   - Break down the load into smaller parts if you can.
2. **Bend the knees.** This is the single most important rule when lifting moderate to heavy objects. Take a tip from professional weight lifters. They can lift tremendous weights because they lift with their legs, not their backs.
   - When lifting a box, position your feet close to it.
   - Center yourself over the load.
   - Bend your knees and get a good hand hold.
   - Lift straight up, smoothly.
   - Allow your legs, not your back, to do the work.

3. **Do not twist or turn your body once you have made the lift.** Keep the load close to your body, and keep it steady. Any sudden twisting or turning could result in taking out your back.
4. **Make sure you can carry the load where you need to go before attempting to move it.** Also, make sure your path is clear of obstacles and that there are no hazards, such as spilled grease or oil in your path. Turn your body by changing foot positions, and have sure footing before setting out.
5. **Set the load down properly.** Setting the load down is just as important as lifting it. Lower the load slowly by bending your knees, letting your legs do most of the work. Don't let go of the load until it is secure on the floor.
6. **Always push, not pull**, the object when possible. When moving an object on rollers, for example, pushing puts less strain on the back and is safer, should the object tip.

Planning Ahead

Planning ahead makes sense. If you know certain loads will have to be carried from storage, place the objects on racks, not on the floor, whenever possible. That way the load will not have to be lifted from the floor. Do not attempt to carry loads that are clearly too heavy for you. Long objects, such as pipes and lumber, may not be heavy, but the weight might not be balanced. Such objects should be carried by two or more people.

If the load can be split up into smaller ones, you're better off doing that, even if loading takes a few extra minutes. Trying to lift it all at once or even in two or three loads may be asking for trouble when the weight is great.

When catching falling or tossed objects, your feet should be firmly planted, with your back straight and your knees slightly bent. Your legs should absorb the impact, not your back. If you're working on something low, bend your knees. Keep your back as straight as possible. Bending from the waist can lead to back pain. In both of these situations, frequent rest breaks are necessary to keep from getting back fatigue.

When a Serious Backache Happens

Although with simple care and bed rest, most back pain goes away, a serious back injury or chronic back pain will require treatment. If the pain does not go away, or is accompanied by weakness or numbness in the lower limbs, you should see your doctor. Pain that radiates from the back to the buttocks and legs is typical of lower back disorders and is called sciatica.

When you go to your doctor, in addition to giving your medical history and having a physical exam, you may need other tests to determine the exact source of the pain. Today, equipment is available to help your physician determine problems.

Treatment may consist of bed rest, cold or hot packs, traction, physical therapy, or muscle-relaxing drugs. Some treatment requires injections around the spinal nerves and, in some cases, surgery may be necessary.

Certain jobs require long hours of standing or sitting. These conditions can create back troubles. Get up and stretch frequently if you are required to sit for long periods. If standing, ease the strain on your lower back by changing foot positions often, placing one foot on a rail or ledge. However, keep your weight evenly balanced when standing. Don't lean to one side.

Work at Working Safely

By using common sense, you can help keep your back out of trouble. Every time you think about lifting, think defensively about your back and the possibility of a back sprain. Follow good lifting techniques, not only at work, but also at home. It's your back and your life.

With proper exercise, a good diet and the proper lifting techniques, your chances of being out of work with chronic or severe back pain are greatly reduced. Remember to:

1. Plan ahead when lifting jobs are necessary.
2. Get help to lift objects that are too heavy for you.
3. Never twist or turn suddenly while carrying a heavy load.
4. Make sure your path is clear and be careful of your footing.
5. Lift with the legs, not your back.
6. Be aware of proper posture when sitting, standing, or reclining.
7. Follow a sensible diet and exercise program to help your back.

Following these simple rules reduces your risk of injury to your back. If you have ever had back pain you know how important this is. If you have not suffered from back pain, following these rules will help assure that you never will.

LOCKOUT/TAGOUT: THE CONTROL OF HAZARDOUS ENERGY

Lockout/tagout procedures are for your safety. They are designed to prevent accidents and injuries caused by the unexpected release of energy. These procedures prevent workers from accidentally being exposed to injurious and even life-threatening situations with energized machinery.

Where Are the Regulations?

The Occupational Safety and Health Administration (OSHA) regulates lockout/tagout through the Control of Hazardous Energy standard, found at 29 CFR 1910.147. This standard mandates training, audits, and recordkeeping to ensure that workers will not be injured by unintentionally energized equipment.

What is Lockout/Tagout?

Lockout is the process of preventing the flow of energy from a power source to a piece of equipment, and keeping it from operating.

Lockout is accomplished by installing a lockout device at the power source so that equipment powered by that source cannot be operated. A lockout device is a lock, block, or chain that keeps a switch, valve, or lever in the off position.

Locks are provided by the employer and can be used only for lockout purposes. They should never be used to lock tool boxes, storage sheds, or other devices.

Tagout is accomplished by placing a tag on the power source. The tag acts as a warning not to restore energy—it is not a physical restraint. Tags must clearly state: **Do not operate** or the like, and must be applied by hand.

Both locks and tags must be strong enough to prevent unauthorized removal and to withstand various environmental conditions.

What Must Be Locked or Tagged Out

The control of hazardous energy standard (lockout/tagout), covers servicing and maintenance of equipment where unexpected energization or start-up of equipment could harm employees.

You need to control energy before working in situations involving repair and replacement work, renovation work, and modifications or other adjustments to powered equipment. There may be other instances as well when lockout/tagout is required at your facility.

In general, OSHA requires that all power sources that can be locked out, must be locked out for servicing or maintenance. Remember, guards or interlock devices cannot be used as substitutes for locks during major servicing.

The standard requires that employers develop written energy control programs that clearly and specifically explain all procedures for lockout/tagout. These plans must include:

- Lockout/tagout procedures.
- Employee training.
- Periodic inspections.

Employers must identify and differentiate between authorized and affected employees. **Authorized employees** physically lock or tag out equipment for servicing or maintenance. Note that these individuals are not necessarily the people who normally operate the equipment.

Affected employees are those workers whose job requires them to operate equipment subject to lockout/tagout, or those employees who work in areas where lockout/tagout is used. Your employer will inform you if you are an affected employee.

Controlling Energy Sources

A wide variety of energy sources require lockout/tagout to protect you from the release of hazardous energy. Some of these energy sources include:

- **Electrical**
- **Mechanical**
- **Pneumatic** (involving gases, especially air)
- **Hydraulic** (involving fluids, especially water)
- **Chemical**
- **Thermal**
- **Water under pressure** (or steam)
- **Gravity**
- **Potential**

Lockout/tagout must be used to protect you from the potentially dangerous effects of hazardous energy. Some of the problems of hazardous energy include:

- **Accidental start-ups**
- **Electric shock**
- **Release of stored, residual, or potential energy**

Remember, these accidents often occur when someone takes a short cut when servicing machinery, or they occur when a worker doesn't understand the equipment or job to be done.

Before the standard went into effect in 1989, OSHA estimated that failure to control hazardous energy sources caused:

- 10 percent of serious industrial accidents.
- 33,000 lost work days each year.
- Loss of about 140 lives each year.

It was because of this serious risk to worker safety that the standard was developed.

The Lockout/Tagout Procedure

Lockout/tagout procedures cover the following:

- The scope and purpose of lockout/tagout.
- How to perform a shutdown, including isolating, blocking, and securing machines or equipment.
- How to place, remove, and transfer locks and who is responsible for them.
- How to test the machine to make sure it is locked out.

Preparing for a Shutdown

Before you ever turn off a machine as part of a lockout/tagout procedure, you should know:

- The type and magnitude of the energy involved.
- Associated hazards of the energy involved.
- Control methods of the energy involved.

Performing a Shutdown

First, notify all affected employees that you're about to start a lockout procedure. Then locate all energy sources that power the piece of equipment you'll be servicing. Always look for hidden energy sources. Some machines may have more than one source of power, so you must make sure you know the machine and all power sources involved. Follow the procedures set up to shut down each respective machine.

Isolating Equipment and Applying Lockout Devices

Your machines or equipment to be locked out should already be capable of being locked out. Every power source has its own procedure for lockout. Lockout may be accomplished by pulling a plug, opening a disconnect switch, removing a fuse, closing a valve, bleeding the line, or placing a block in the equipment. Generally, follow this sequence of events:

- After you have completed the shutdown, turn off the energy at the main power source.
- Using your designated lock, lock out all energy sources involved.
- Attempt to restart the machine to guarantee that the power is shut off, then return the switch to the off position.

If several people are needed to work on a piece of equipment, each one must apply their own lock. This prevents any accidental start-ups while another employee may still be working on the machinery. In this case, you'll use a **multiple lockout device** that can accommodate several locks at once.

When all energy sources are locked, inform others of the lockout situation. One way to do this is by applying a tag to the power source.

Note: Never use another employee's lock and never lend yours. This protects you and your fellow workers.

Safe Release of Stored Energy

Equipment must be at "zero energy state" before servicing or maintenance work can begin. To get to this zero energy state:

- Drain all valves, bleed off air from a system, eliminate stored hydraulic pressure, or use any method to release energy that is detailed in your company procedure.

- Test the machine to make sure that all energy was disconnected or released.

Verify That Machine Is Locked Out

Before you start to repair or service the machine, make sure that it has been properly isolated and deenergized. With your lock in place, test the disconnect to make sure it can't be turned on. Make absolutely sure the power can't be supplied unless you know about it.

Restoring Power

After servicing is finished, check that all tools are removed from the area and replace all machine guards. Make sure all employees are clear of the machine. Only then can you remove your tag and lock and reconnect all sources of energy. After this, you may notify the affected employees that the lockout has been removed and restart the equipment.

Following Training and Audit Requirements

OSHA requires that:

- All **authorized employees** be trained in recognition of applicable hazardous energy sources, the type and magnitude of hazardous energy sources in use at the facility, and how to perform the lockout/tagout procedure.

- All **affected employees** must be trained in the purpose and use of lockout/tagout.

- All other employees must be instructed on the purpose of the plan, but not in the actual use.

- Periodic inspections or audits be performed by an authorized employee who does not use the energy control procedure being inspected.

- Retraining must be done when there are changes in equipment, job assignment, or procedures, when an audit shows deficiencies with the procedure, and when the employer feels the procedures should be reviewed.

Audits must be done at least annually, and should include questions to determine if employees understand the purpose of lockout/tagout, if proper locks and tags are being used, and if established procedures are being followed. Each audit must be documented.

Other Concerns

Other concerns that must be addressed for your organization's lockout/tagout program include working with outside contractors, shift and personnel changes, and power sources that cannot be locked out.

Outside Contractors

Outside contractors must be informed of your lockout/tagout procedure in full detail so that their employees understand the meaning of locks or tags that they may come across during the course of their work. In addition, if contractors will be using locks or tags, they should inform your employer so that everyone affected may be notified.

Shift and Personnel Changes

In general, if a piece of equipment is locked out at shift change, the person on the next shift must apply his lock before the employee who is leaving can remove his.

Power Sources That Cannot Be Locked Out

In very rare cases, a power source cannot be physically locked out. Discuss this situation with your supervisor to find out if tagout alone may safely be used. There are a few situations where tagout alone is allowed.

Work at Working Safely

Your attention to and respect for your facility's lockout/tagout program will make the workplace safer for both you and your co-workers. Always follow lockout/tagout procedure during servicing or maintenance of equipment, where unexpected energization or start-up of the equipment could harm you or a fellow employee.

1. Always lock and tag out power sources and switches when you service or repair energized equipment.
2. Never ignore or remove the locks or tags of other employees when you come across them in the workplace.
3. Know your role as an authorized or affected employee.

MACHINE GUARDING: WORKING SAFELY WITH MACHINES

While machines allow more efficient, productive work, you must use them with great caution. Safety should be foremost in your mind when working with moving machine parts. It's up to you to wear protective equipment, maintain equipment, and use safety features and tools correctly. You are in charge of your own personal safety on the job.

Where Are the Regulations?

The Occupational Safety and Health Administration (OSHA) has put forth several regulations that apply to the use of electrically powered machinery, under Subpart O, Machinery and Machine Guarding, and Subpart P, Hand and Portable-Powered Tools and other Hand-Held Equipment. These rules can be found in 29 CFR 1910.211-.247.

There are also guarding requirements under resistance welding 29 CFR 1910.255(a)(4) and (b)(4). These requirements touch on lockout/tagout procedures during welding operations and point of operation guards for press welding machines.

In general, remember that any machine part, function, or process which may cause injury must be guarded. Where the operation of a machine or accidental contact with it, can injure you or others, the hazard must be either controlled or eliminated.

Serious Injuries Are Possible

Crushed hands and arms, severed fingers, blindness—the list of possible machinery-related injuries is as long as it is horrifying. There seem to be as many hazards created by moving machine parts as there are types of machines. Guards are essential for protecting workers from needless and preventable injuries.

In addition, most machines and power tools are powered by electricity. Electrical hazards are equally debilitating. Electricity will give you a shock if you accidentally become a ground. Breathing can stop and nerve centers may be temporarily paralyzed. Your heart beat is interrupted so blood stops circulating. Heat from the current can cause internal bleeding and destruction of nerves or muscles. The severity of injury depends on where current flows and how long, not the voltage. For example, did you know that $^{60}/_{1000}$ of an ampere can kill you if it passes through the chest?

You can see that it's absolutely necessary to pay attention as you use equipment. A machine can be pretty unforgiving if you slip up—be sure you're in charge.

Where Mechanical Hazards Occur

These types of dangerous moving parts need guarding:

- **The point of operation**, or that point where work is performed on the material, such as cutting, shaping, boring, or forming of stock.

- **Power transmission apparatus**, or the components of the mechanical system which transmit energy to the part of the machine performing the work. These components include flywheels, pulleys, belts, connecting rods, couplings, cams, spindles, chains, cranks, and gears.

- **Other moving parts**, or parts of the machine which move while the machine is working, can include reciprocating, rotating, and transverse moving parts, as well as feed mechanisms and auxiliary parts of the machine.

Hazardous Mechanical Motions and Actions

Different types of hazardous mechanical motions and actions are basic to nearly all machines. Recognizing them is the first step you can take toward protecting yourself from the dangers they present. We will briefly examine the following types of hazards in turn.

Rotating motion can be dangerous; even smooth, slowly rotating shafts can grip clothing, and through mere skin contact force an arm or hand into a dangerous position. Injuries due to contact with rotating parts can be severe.

Collars, couplings, cams, clutches, flywheels, shaft ends, spindles, and horizontal or vertical shafting are some examples of common rotating mechanisms which may be hazardous. There is added danger when bolts, nicks, abrasions, and projecting keys or set screws are expcsed on rotaiing parts on machinery.

In-running nip points, or those locations that can capture body parts in rotating machinery parts, are common, but dangerous hazards for the machine operator. There are three main types of in-running nips:

- Parts that rotate in opposite directions while their axes are parallel to each other. These parts may be in contact or in close proximity to each other. In the latter case, the stock feed between the rolls produces the nip points. This danger is common on machinery with intermeshing gears, rolling mills, and calenders.
- Another type of nip point is created between rotating and tangentially moving parts. Some examples would be the point of contact between a power transmission belt and its pulley, a chain and sprocket, or a rack and pinion.
- Nip points can also occur between rotating and fixed parts which create a shearing, crushing, or abrading action, for example, spoked handwheels or flywheels.

Reciprocating motions may be hazardous because, during the back-and-forth or up-and-down motion, you might get struck by or caught between a moving and stationary part.

Transverse motion (movement in a straight, continuous line) creates a hazard because a worker may be struck or caught in a pinch or shear point by a moving part.

Cutting action involves rotating, reciprocating, or transverse motion. The danger of cutting action exists at the point of operation where finger, head, and arm injuries can occur and where flying chips or scrap material can strike the eyes or face. Such hazards are present at the point of operation in cutting wood, metal, or other materials. Typical machines having cutting hazards include bandsaws, circular saws, boring or drilling machines, turning machines (lathes), or milling machines.

Punching action results when power is applied to a slide (ram) for blanking, drawing, or stamping metal or other materials. The danger of this type of action occurs at the point of operation where stock is inserted, held, and withdrawn by hand. Typical machinery used for punching operations are power presses and iron workers.

Shearing action involves applying power to a slide or knife in order to trim or shear metal or other materials. A hazard occurs at the point of operation where stock is actually inserted, held, and withdrawn. Machinery used for shearing operations includes mechanically, hydraulically, or pneumatically powered shears.

Bending action results when power is applied to a slide in order to draw or stamp metal or other materials. A hazard occurs at the point of operation where stock is inserted, held, and withdrawn. Power presses, press brakes, and tubing benders all use bending action.

Guard Requirements

What must a guard do to protect you from mechanical hazards? Guards must meet these minimum general requirements:

- **Prevent contact:** The guard must prevent hands, arms, or any part of your body or clothing from making contact with dangerous moving parts.
- **Secure:** Guards should not be easy to remove or alter; a guard that can easily be made ineffective is no guard at all. Guards and safety devices should be made of durable material that will withstand the conditions of normal use. They must be firmly secured to the machine.

- **Protect from falling objects:** The guard should ensure that no objects can fall into moving parts. A small tool which is dropped into a cycling machine could easily become a projectile that could strike and injure someone.
- **Create no new hazards:** A guard defeats its own purpose if it creates a hazard of its own such as shear point, a jagged edge, or an unfinished surface which can cause a laceration. The edges of guards, for instance, should be rolled or bolted in such a way that they eliminate sharp edges.
- **Create no interference:** You might soon override or disregard any guard which keeps you from doing your job quickly and comfortably. Proper guarding can actually enhance efficiency since it can relieve your worries about injury. If possible, one should be able to lubricate the machine without removing the guards.

MACHINE GUARDING

Even the most elaborate guarding system cannot offer effective protection unless you know how and why to use it. You should be aware of the following:

- A description and identification of the hazards associated with particular machines.
- The guards themselves, how they provide protection, and the hazards for which they are intended.
- How to use the guards and why.
- How and under what circumstances guards can be removed, and by whom (in most cases, repair or maintenance personnel only).
- What to do (e.g. contact your supervisor) if a guard is damaged, missing, or unable to provide adequate protection.

Machine Guarding Methods

There are many ways to guard machinery. The type of operation, size or shape of stock, method of handling, physical layout of the work area, type of material, and production requirements or limitations will help to determine the appropriate method for a given machine.

As a general rule, power transmission apparatus is best protected by fixed guards that enclose the danger area. For hazards at the point of operation, where moving parts actually perform work on stock, several kinds of guarding are possible. Guards can be grouped under five general categories:

Guards are barriers which prevent access to danger areas.

A safety **device** may perform one of several functions. It may:

- Stop the machine if a hand or any part of the body is inadvertently placed in the danger area.
- Restrain or withdraw the operator's hands from the danger area during operation.
- Require the operator to use both hands on machine controls, thus keeping both hands and body out of danger.
- Provide a barrier which is synchronized with the operating cycle of the machine in order to prevent entry to the danger area during the hazardous part of the cycle.

Guarding by location or distance has many applications. A thorough hazard analysis of each machine and situation is necessary before attempting this technique.

The machine or its dangerous moving parts must be positioned so that hazardous areas are not accessible or do not present a hazard during the normal machine operation to guard a machine by location. For example, locating a machine so that a wall protects the worker is guarding by location.

Feeding and ejection methods of guarding limit hazards associated with feeding stock into machines once it starts to function.

MACHINE GUARDING

Miscellaneous aids do not provide complete protection from machine hazards, but provide an extra margin of safety. One example is an awareness barrier. An awareness barrier serves to remind you that you are approaching a danger area.

Personal Protective Equipment

Engineering controls that eliminate the hazard at the source and do not rely on behavior for their effectiveness offer the best and most reliable means of safeguarding. Therefore, engineering controls must be the employer's first choice for eliminating machine hazards. But whenever engineering controls are not available or are not fully capable of protecting you, you must wear protective clothing or personal protective equipment (PPE).

PPE is, of course, available for different parts of the body. Hard hats can protect the head from the impact of bumps and falling objects when you work with stock. Caps and hair nets can help keep your hair from being caught in machinery. If machine coolants could splash or particles could fly into the operator's eyes or face, then face shields, safety goggles, glasses, or similar kinds of protection might be necessary. Hearing protection may be needed when operating noisy machines.

To guard the trunk of the body from cuts or impacts from heavy or rough–edged stock, there are certain protective coveralls, jackets, vests, aprons, and full–body suits. Workers can protect their hands and arms from the same kinds of injury with special sleeves and gloves. Safety shoes and boots, or other acceptable foot guards, can shield the feet against injury when handling heavy loads which might drop.

It is important to note that protective clothing and equipment can create hazards. A protective glove which can become caught between rotating parts, or a respirator facepiece which hinders the wearer's vision, for example, require alertness and continued attentiveness whenever they are used.

Other clothing may present additional safety hazards. For example, loose-fitting shirts might possibly become entangled in rotating spindles or other kinds of moving machinery. Jewelry, such as bracelets and rings, can catch on machine parts or stock and lead to serious injury by pulling a hand into the danger area.

Some General Safety Rules

General safety rules apply to both stationary and portable equipment. Never let overconfidence lead you into taking unnecessary risks. The following rules apply to every machine or power tool you use:

- Keep your work area well lit and dry.
- Maintain your tools. For best and safest performance, keep them sharp, oiled and stored in a safe, dry place. Regularly inspect tools, cords and accessories. Repair or replace problem equipment immediately.

- Keep your work area clean. Sawdust, paper, and oily rags are a fire hazard and can damage your tools.
- Use safety features like three-prong plugs, double-insulated tools, and safety switches. Make sure machine guards are in place on large and small equipment.

- Use protective equipment when necessary. This might include safety glasses, hearing protection and respiratory protection.

- Dress right. Never wear clothing or jewelry that could become entangled in power tools.

- Install or repair equipment only if you're qualified. A faulty job may cause fires or seriously injure you or other workers.

- Use the right tool for the job. Don't force a small tool to do heavy-duty work.

- Keep electric cables and cords clean, free from kinks. Never carry a tool by its cord.

Good tool habits soon become second nature. Follow the machine safety guidelines at your workplace and the equipment you operate will serve you efficiently and safely.

Grounding Is an Important Precaution

Grounding is one of the most important safety measures to take when working with electric equipment. It provides a safe path for electricity, preventing leakage of current in circuits and equipment. Grounding should be provided for the entire system and individual pieces of equipment. Check all ground connections regularly for tightness.

Portable Power Tools

Saws

The circular saw is a heavy-duty tool with interchangeable blades for all types of woodcutting. The saber saw is somewhat smaller and used for smaller woodcutting jobs and curved cuts. A chainsaw may be either gasoline or electrically powered. Follow these safety rules when using saws:

- Before cutting, inspect the material to be cut for nails or foreign objects.
- Make sure blade guards are in place and working properly.
- Stay alert! Saws are noisy and the sound may drown out warning shouts or instructions.

- Wear goggles or goggles and a face shield to protect yourself from flying debris or sawdust.
- Inspect blade regularly. First, turn the saw off and unplug it. Don't use dull or loose blades.
- Don't overload the motor by pushing too hard or cutting material that is too heavy.

MACHINE GUARDING

- Be sure you have firm footing and balance when using any saw. Slips or falls can be deadly when you're holding a power tool.

Portable Drills

Variable speed drills are versatile tools used for boring holes, turning screws, buffing, and grinding. Keep these pointers in mind when using them:

- Select the correct drill bit for the job to be done. Use only sharp bits.

- Make sure the material being drilled is secured or clamped firmly.

- Hold the drill firmly and at the correct angle. Don't force it to work or lean on it with all your strength.

- Always remove the bit from the drill when you're finished.

For restoring a cutting edge to drill bits, use a drill bit sharpener. It should be double-insulated and placed flat on a bench surface. Don't forget to wear safety glasses when you use the sharpener.

Grinding Wheels

Bench grinders are useful for sharpening, shaping, and smoothing metal, wood, plastic, or stone:

- Keep machine guards in place and wear ear and eye protection.

- Before use, make sure that wheels are firmly held on spindles and work rests are tight.

- Stand to one side while starting the motor, until operating speed is reached – this prevents injury if a defective wheel breaks apart.

- Use light pressure when starting grinding, too much on a cold wheel may cause failure.

Portable Sanders

These tools make finishing work faster. Two types are orbital and belt. Remember these tips:

- Arrange the cord so that it won't be damaged by the abrasive belt.

- Keep both hands on the tool for good control.
- Hold onto the sander when you plug it in.

MACHINE GUARDING

- Clean dust and chips from the motor and vent holes regularly and lubricate when necessary.

Miscellaneous Portable Tools

Impact Wrenches — They operate on electricity or compressed air and deliver extra power and torque for fastening and loosening bolts and drilling. Don't force a wrench to take on a job bigger than it's designed to handle. Don't use standard hand sockets or driver parts with an impact tool, they can't take the sharp blows. Don't reverse direction of rotation while the trigger is depressed.

Soldering Irons or "Guns" — They can be dangerous because of the heat they generate. Handle with care – they can easily cause third degree burns. Always assume that a soldering iron is hot. Rest a heated iron on a rack or metal surface. Never swing an iron to remove solder. Hold small soldering jobs with pliers, never in your hand. When cool, store it in its assigned area.

Propane and Gas Torches — These commonly used tools pose flame and heat hazards. Never use a flame to test for propane or gas leaks.

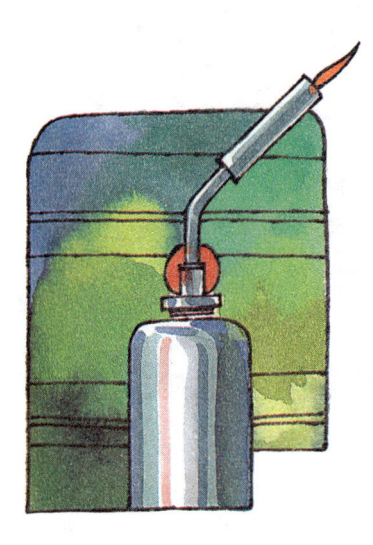

Never store the fuel tanks in an unventilated area and never use a tank with a leaking valve. Use torches in well-ventilated areas. Avoid breathing the vapors and fumes they generate.

Glue Guns — A glue gun can be a real time saver. However, because it generates temperatures as high as 450°F, avoid contact with the hot nozzle and glue.

Shop Vacuums — They enable you to keep a safe, clean work place. Use the correct hose size and accessory for the job you're doing. Clean filters regularly and never use your vacuum to pick up flammable liquids or smoldering materials.

Safety Rules for Stationary Machinery

These are the big workhorses of the shop and plant. Remember to always stay alert and work with caution. These tools are powerful and often more complicated than their smaller cousins. First, a few general rules that apply to operating machines:

- Use all guards and safety devices that are designed to be used with the equipment.
- Never use a dull blade or cutting edge.

- Make adjustments and accessory changes when machinery is turned off and unplugged.
- If you're tired, take a break. Also, don't take your eyes off your work or talk to anyone as you use the tools.
- Dress right, don't wear loose fitting clothing that can get caught.

Table Saw

This saw has a large circular blade used to make a variety of cuts in wood or other material:

- Never reach over the saw to push stock that has been sawed.

- Stand slightly to one side, never in line with the saw.

A "kickback" occurs when material being cut is thrown back toward the operator. This is one of the greatest hazards in running a table saw. To avoid it:

- Never use a dull blade.
- Don't cut "freehand" or attempt to rip badly warped wood.
- Use the splitter guard.
- Don't drop wood on an unguarded saw.

Radial-Arm Saw

Often called the number one multipurpose saw in the shop, this saw blade is mounted on a moveable head, and slides in tracks or along a shaft. Most have built-in safety devices such as key switches to start them, blade guards, anti-kickback pawls, and blade brakes. Follow these precautions:

- The saw and motor should always be returned to the rear of the table against the column after a cut is made.
- If the motor slows while cutting, it means it is overloaded. This can be due to low voltage, bad blades, or material being fed too fast.
- Keep the machine in good alignment and adjustment to prevent excessive vibration.

Drill Press

The stationary drill press is a larger, more powerful version of a portable drill. Remember to:

- Clamp or securely fasten the material being drilled whenever possible.
- Make sure any attachments are fastened tightly.

Miscellaneous Stationary Tools

Power Sanders. These machines do finishing work in a fraction of the time it would take by hand. Always select the correct grade of abrasive for the job. Move the work around to avoid heating and burning a portion of the disk, belt, or wood. Remember to use the dust collector if the sander has one.

Shapers. A shaper is used mainly for grooving and fluting woods. It can be dangerous because of its high speed and because the cutters are difficult to guard completely. When using a shaper, avoid loose clothing, wear eye protection and make sure the cutters are sharp and securely fastened.

Welding Machines. The high-intensity arc of even a small welding machine can cause severe burns. Non-flammable clothing and hand and eye protection are needed to protect against hot sparks and molten metal. Keep the area around the welding operation clean—hot sparks can start fires.

Work at Working Safely

Proper care and safety when using machinery is vital.

1. Respect your equipment, know the dangers it presents, and take safety precautions necessary to work without injury.
2. Maintain equipment with regular servicing and good housekeeping practices.
3. If you don't know how to use a particular piece of equipment, don't be afraid to admit it. Find someone who does and learn from an experienced worker.
4. Think safety on the job to ensure that you and your equipment will have a long and productive life.

PERSONAL PROTECTIVE EQUIPMENT: DO YOUR SHARE TO AVOID INJURY

According to the Bureau of Labor Statistics (1997), 2.8 million people suffered serious, nonfatal on-the-job injuries and illnesses. That works out to an average of over 7,000 persons injured per day throughout the year. In addition, an average of 17 American workers die each day from injuries sustained on the job (1996).

These sobering statistics demonstrate that many workers face unsafe conditions or work practices in their workplaces every day. While employers need to minimize these hazards as much as possible at the source, this step is not always feasible. Use of personal protective equipment (PPE) completes other measures your employer takes to create a safe work environment for you.

Where Are the Regulations?

The Occupational Safety and Health Administration (OSHA) has issued regulations governing the use of PPE in general industry. You can find them at 29 CFR 1910.132-.138. Your employer must establish and administer an effective PPE program.

Hazard Assessment and Equipment Selection

Your employer must assess your workplace to determine if hazards are present, or are likely to be present, which necessitate the use of PPE.

OSHA does not want your employer to rely only on PPE to protect against hazards, but rather to use PPE along with guards, engineering controls, and sound manufacturing practices.

Good Communication is a Top Priority

Safe workplaces, free of unnecessary hazards, are made possible through careful planning, preparation, and good communication. When your employer gets ready to survey your workplace, find out how you can contribute. Be ready to explain each step of your job and point out potential job hazards.

Your employer will note situations where PPE is currently used, what type, and for what purpose. He or she will be looking for hazards like:

1. Sources of motion – machinery or processes where any movement of tools, machine elements, or particles could exist; or movement of personnel that could result in collision with stationery objects.

2. Sources of high temperatures – could result in burns, eye injury, or ignition of protective equipment.

3. Types of chemical exposure – handling of chemicals during a production process, or the exposure from a possible spill or leak.

4. Sources of harmful dust – areas where cutting metal, concrete, or other operations produces dust.

5. Sources of light radiation – welding, brazing, cutting, furnaces, heat treating, high intensity lights, etc.

6. Sources of falling objects or potential for dropping objects – manlifts in warehousing, stacked pallets, using dollies, shipping areas.

7. Sources of sharp objects which might pierce the feet or cut the hands – working with machinery, food handling and storage, sawing and cutting.

8. Sources of rolling or pinching objects which could crush the feet – moving stock, such as paper rolls.

9. Any electrical hazards.

PPE

10. Co-workers – people who work in the immediate vicinity of others can present hazards from their presence or the operations they are involved with.

Your employer has many reasons for performing a hazard assessment. An assessment can point out areas of high accidents and injuries, enabling process changes. It can tell where tools or equipment need to be repaired or replaced, before an accident occurs. An assessment can identify outdated or inefficient work practices.

Your Responsibility to Report Hazards

Think about the potential for a hazardous situation in your workplace. Would you know what to do to respond effectively? How do you report a safety problem or hazard?

A hazard assessment reflects an employer's responsibility to provide a workplace free of recognized hazards that are likely to cause death or serious harm. Similarly, you have a responsibility to report hazards you discover. In addition to participating in hazard assessments, make these common sense rules part of your routine on the job:

- Identify all potential hazards before you begin a task.
- Respect all precautions – don't take any chances.
- Check with a supervisor or somebody else in authority if you are unsure about a situation.
- Know in advance the potential problems in a situation, and what to do about them if they happen.
- Know your organization's hazard reporting procedure.
- Learn basic first-aid procedures and use them on the job only if your employer approves.
- Report any hazards to a supervisor or designated person as soon as you become aware of them.

Under the General Duty Clause (Sec. 5(a)(1)) of the OSH Act, you have the right to safety and health on the job. Sec. 11(c) guarantees that right without fear of punishment or reprisal from your employer.

Work at Working Safely

Because your health is important to you **and** your employer, take the following points to heart:

- Use common sense regarding safety on the job and comply with any applicable OSHA standards.
- Work with your employer in identifying hazards on the job.
- Report any job-related injury or illness promptly, and seek recommended treatment.
- Follow your employer's safety and health rules and regulations, including the use of personal protective equipment on the job.

PPE

The goal of hazard reporting should be to make the workplace a safer environment for all employees. That goal needs everybody's support.

Eye Protection: Seeing Is Believing

When it comes to eye protection, you and your employer share responsibility for your safety. Your employer can provide safety equipment, first-aid facilities, and even a vision screening program, but YOU have to take safety seriously and use these protections.

Where Are the Regulations?

Regulations have been issued by the Occupational Safety and Health Administration (OSHA) on personal protective equipment (PPE) in general, and on eye protection in particular. You can find these regulations at 29 CFR 1910.132-.133. These regulations require employees to use eye protection to guard against injury in situations where reasonable probability of injury exists.

You must receive training in the proper use of eye and face protection and understand the following concepts:

- When eye and face protection is necessary.
- What eye and face protection is necessary.
- How to properly put on, take off, adjust, and wear goggles, face shields, etc.
- PPE limitations.
- Proper care, maintenance, useful life, and disposal of eye and face protection.

You must show that you understand the training, and can use eye and face protection properly before you will be allowed to perform work requiring its use.

Eye Injuries Are Often Permanent

An eye injury resulting in blindness cannot be cured. Excuses like "I don't wear my goggles because my hair gets messed up" or "I look silly in safety glasses" seem unimportant when compared with the value of a pair of healthy eyes. Proper eye protection reduces your chances of injury and reduces the severity of injury if an accident does occur.

An old safety proverb makes this point well, "You can walk with a wooden leg, you can chew with false teeth, but you can't see with a glass eye!"

How Do Injuries Happen?

OSHA reported 66,000 disabling eye injuries in 1996. The main cause of job-related eye injuries is objects striking a worker's eye. Chemical splashes also account for many eye injuries.

What Are the Hazards?

Most workers who had eye injuries were not wearing eye protection. They said that eye protection was not normally used or they felt it wasn't needed.

Eye injuries can be avoided by following safety precautions and wearing proper protective equipment. Here are some of the causes of eye injuries in detail, and some workplace operations where they are often found.

Injurious gases, vapors, and liquids. Workers handling acids or caustics, and doing welding are subject to these hazards.

Dusts or powders, fumes and mists. Some sources are scaling, light grinding, spot welding, and woodworking; they can also include very small flying particles.

Flying objects or particles. Some sources include, chiseling, grinding, hammering, and metal working. These hazards cause the majority of eye injuries.

Splashing metal. Some sources are babbitting, casting of hot metal, and dripping in hot metal baths.

Thermal and radiation hazards such as heat, glare, ultraviolet, and infrared rays. Sample sources are welding, metal cutting, and furnace tending.

Lasers. A recent addition to the list of eye hazards, laser beams can present dangerous and unusual exposure. Different kinds of laser beams require different methods of eye protection.

Electrical hazards. Sample sources are arcing and sparks.

How Can You Protect Your Eyes?

Your employer must identify hazards in your work area to determine how best to avoid eye injuries. The first steps to prevent eye injuries are to reduce the occurrence of foreign objects, install equipment guards, and provide PPE. Your employer must also provide eyewashes to minimize damage once an injury has occurred.

Equipment Guards

Plant equipment and machinery is the source of many eye injuries. Be sure to use any guards, screens, and shields that are attached to equipment. Make sure they are always in place and used along with additional eye protection.

Movable screens are available for work settings like machine shops. The screens can be used to separate workers at one lathe from those at nearby work stations. Portable welding screens can be positioned around welding areas to protect other workers from sparks and radiation.

Ventilation and Lighting

It is also important that work areas have good lighting and ventilation. Proper lighting reduces glare and eye strain and enables you to see your work clearly. A good ventilation system will carry away flying debris that might be hazardous to the eyes if it remains in the atmosphere.

Eyewash Facilities Are Important

No one can predict when and where an accident will occur. Therefore, you should be familiar with the location and operation of emergency eyewash facilities. These can include eyewash fountains, drench showers, hand-held drench hoses and emergency bottles. Very simply, they all use large amounts of water to flush away eye contaminants.

Location Is Important

The location of eyewash facilities is very important because your eyes can be damaged very quickly by many contaminants. The first fifteen seconds after the injury is the critical period. Because of this critical time period, the American National Standards Institute (ANSI), suggests that eyewashes be within 100 feet or a 10 second walk of the work area (ANSI Z358.1-1990).

Eyewashes should not be installed where workers would have to pass through a doorway, go up or down stairs, or weave between equipment to get help. If you get something in your eye—dirt, wood, metal, or a flying particle—go immediately to the nearest eyewash.

What to Do

Flush the eye with water until the foreign object has been rinsed out. Don't rub your eye, this can scratch the eye or embed the object. If you can't rinse out the object, bandage your eye loosely and get additional medical attention.

If a chemical splashes in your eye, move quickly to an emergency shower or eyewash. Look directly into the stream of water and hold your eye open with your fingers. Flush your eye for at least 15 minutes and then get first aid.

Practice Makes Perfect

It's a good idea to practice using the eyewash and to become familiar with how it works. You might even practice holding your eyes open in a stream of water. It's a natural reaction to squeeze the eyes closed tightly when you get something in them. This reaction might prevent you from washing out your eyes quickly in case of an emergency.

Personal Protective Equipment

A wide variety of safety equipment is available to keep you safe and injury free. The safety devices and procedures listed below are all ways to ensure eye protection and continued eye health. Protective eye and face equipment must comply with ANSI guidelines and be marked directly on the piece of equipment (e.g. glasses frames and lenses).

Safety Glasses

The most common type of protective equipment for the eyes is safety glasses. They may look like normal street-wear glasses; they are made of glass, plastic, or polycarbonate. But, they are made much stronger than street-wear lenses, are impact resistant, and come in prescription or nonprescription (plano) forms.

Polycarbonate lenses are light and offer greater impact resistance, but glass lenses are especially good protection against infrared radiation. So the choice of lens material will depend on your work situation. Tinted lenses and anti-glare protection are available for daylight use as well as nighttime situations where very bright lights are present. There are also special coatings available which prevent fogging of lenses.

Safety frames are stronger than street-wear frames and are heat-resistant. They also help prevent lenses from being pushed into your eyes. Different styles of frames are available for different jobs.

Safety glasses also are available with side shield guards. Semi-side shields provide protection for the sides of your eyes. Eye-cup side shields provide more thorough eye protection from hazards that come from the front, side, top, or bottom.

Goggles

Goggles are very similar to safety glasses but fit closer to the eyes. They can provide additional protection in hazardous situations involving liquid splashes, fumes, vapors, and dust. Some models can be worn over prescription glasses.

You should maintain and clean your safety glasses and goggles regularly. Dirty, scratched, or cracked lenses reduce vision and seriously reduce protection. Replace damaged glasses immediately.

Face Shields

Full-face protection is often required to guard against molten metal and chemical splashes. Face shields are available to fit over a hard hat or to wear directly on the head. A face shield should always be used with other eye protection such as goggles or glasses.

What about Contact Lenses?

Most workers can safely wear their contacts on the job. Situations where contacts should be worn with caution include workplaces where you might be exposed to chemical fumes, vapors or splashes, intense heat, and molten metals.

It is important to remember that, if hazards warrant, your contacts should be worn along with additional eye protection. Contacts should be removed immediately if redness of the eye, blurring of vision, or pain develops on the job.

It's also a good idea to keep a spare pair of contacts or prescription glasses with you in case the pair you usually wear is lost or damaged while you're working. You might also want to make sure your supervisor or plant first-aid personnel know that you wear contacts, in the event of any injury on the job.

When Are Absorptive Lenses Required?

You have probably experienced the eye fatigue that results from a long day of driving in the bright sun. It is now believed that this glare can actually seriously injure the human eye. As an industrial worker, you may be exposed to extreme light conditions and glare from the sun, bright lights, welding, brazing, or soldering processes.

Absorptive lenses are used to absorb or screen out unwanted light and glare. Most ordinary sunglasses do not provide the right glare protection. For welding or work with torches, goggles or helmets are available with filter lenses to shield the eyes from radiation and glare.

Don't Ignore Vision Testing

Uncorrected vision problems can contribute to accidents that lead to eye injury and loss of sight. Eye strain, double vision, and lack of depth perception can reduce your efficiency and productivity. Some employers now offer vision screening tests. These tests are sometimes used to match individuals to jobs requiring specific vision skills. Not everyone's sight is right for every job.

What It Looks for

Vision testing checks the sharpness of vision, muscle balance, and color distinguishing capability. Last but not least, the screening process is able to detect health problems such as glaucoma, diabetes, or cataracts. Vision screening frequency depends on the vision requirements of the job and the age of the employee. For instance, crane operators should be tested each year because they need excellent distance vision.

Take advantage of any vision screening program offered by your employer to protect your eyes on and off the job. Some employers even pay for prescription safety glasses.

If a vision test shows that you need prescription glasses, get them! If corrective lenses are called for, it is very important that they are adjusted properly by a professional. The lenses must be correctly positioned in front of the eyes and frames should be adjusted so they don't pinch or bind.

Work at Working Safely

To sum up, you are ultimately responsible for the protection of your eyes. Realistically speaking, you have the most to lose if you don't follow good eye safety practices. Let's review the following important rules about eye safety:

1. Match safety equipment to the degree of hazard present.
2. Know what protective devices are available on the job and how they can protect you.
3. Make sure equipment guards are in place on plant machinery and that they are used with additional eye protection.

4. Know location and operation of emergency eyewashes.
5. Inspect eyewashes and showers frequently to make sure they work effectively and that the water is potable.
6. Faceshields should not be used alone, but always with other eye protection such as goggles or glasses.
7. Street-wear eyeglasses are not designed to be safety glasses and should never be used as such.
8. Make sure any safety device you use fits properly.
9. Safety equipment should be maintained in good condition and replaced when defective.
10. Have your eyes tested regularly. If you need corrective lenses, get them and use them!

The goal of eye safety is to protect two of your most valuable possessions—your eyes. The pair you were born with are the eyes that have to last you a lifetime. Protect them!

Fall Protection: Knowledge Is Standard Equipment

You probably spend most of your time at work walking on level surfaces such as office floors, hallways, or the floors of shops and factories. While slips, trips, and falls are still common in such situations, the likelihood of a major injury from such a fall is not great.

However, many employees in general industry work on scaffolds, climb up and down ladders, walk on narrow stairs, work in areas where there may be holes in the floor, or work on elevated floors which have unprotected edges.

Slips, trips, and falls from ladders, scaffolds, stairs, elevated floors and similar work surfaces are major causes of employee injury. In 1996 alone, 317,900 workers suffered injuries from falls which required time off from work.

FALL

Where Are the Regulations?

Although the general industry standards of the Occupational Safety and Health Administration (OSHA) do not yet have fall protection requirements under Subpart I, Personal Protective Equipment (PPE), fall protection requirements exist at:

- Subpart D, Walking/Working Surfaces.
- Subpart F, Powered Platforms, Manlifts, and Vehicle-Mounted Work Platforms.
- Subpart R, Special Industries.

The American National Standards Institute (ANSI) has a consensus standard on *Safety Requirements for Personal Fall Arrest Systems, Subsystems and Components* (ANSI Z359.1-1992). OSHA has issued revised fall protection standards for construction.

Regardless of the current status of any regulations, many employers recognize the need to protect their workers from falls and have set up fall protection programs even though they may not be required. It's a good idea to follow any fall protection guidelines your employer has set up.

Depending on the specific circumstances, your employer may identify workplace fall hazards to which you and your co-workers may be exposed and take steps to minimize or eliminate the hazard. You, in turn, are responsible for following the policies, procedures, and training requirements regarding fall protection your employer establishes.

What Is Fall Protection?

Sooner or later, most people experience that sudden, unanticipated descent in space driven by gravity — a fall. Fall protection encompasses the many means devised to protect workers from fall hazards on the job.

All fall protection systems serve one of two basic functions. They:

- Prevent or restrain a worker from falling.
- Safely stop or arrest a worker who falls.

Guardrail, safety net, and personal fall arrest systems are conventional fall protection systems. They have the widest range of applications and satisfy protection requirements for most tasks that expose workers to fall hazards.

Before you begin using personal fall protection equipment, become familiar with the fall protection systems your employer has put in place to *prevent* falls.

Guardrail Systems and Toeboards

A guardrail is a vertical barrier, normally consisting of an assembly of toprails, midrails, and posts, erected to prevent employees from falling to lower levels. A toeboard is a barrier placed to prevent the fall of materials to a lower level, or to keep employees' feet from slipping over an edge.

Handrail and Stair Rail Systems

A handrail is used to assist employees going up or down stairways, ramps or other walking/working surfaces by providing a handhold for support. A stair rail protects employees from falling over the edge of an open-sided stairway.

Designated Areas

This term refers to a space which has a perimeter barrier erected to warn employees when they approach an unprotected side or edge, and serves also to designate an area where work may be performed without additional fall protection.

Hole Covers

Hole covers, guarding floor openings of at least 2 inches in size, must be capable of supporting the maximum intended load.

Safety Net Systems

Safety nets are non-rigid barriers supported in such a manner as to catch employees who have fallen off a work surface and bring them to a stop before contacting surfaces or structures below.

Safety net systems are conventional arrest systems consisting of mesh nets, including panels, connectors, and other impact-absorbing components. Safety net systems must be installed as close as possible below the surface on which persons are working, but in no case more than 30 feet below the working surface.

Safety nets must be drop tested at the job site after initial installation, relocation, repair, and at six-month intervals if they are left in one place. The drop test consists of a 400-pound bag of sand 30 inches in diameter dropped to the nets from surfaces from which workers could fall.

The maximum size of each safety net mesh opening must not exceed 36 square inches, no longer than six inches on any side or center-to-center.

Ladder Cages

Ladder cages are barriers surrounding or nearly surrounding the climbing area of a ladder. It fastens to the ladder's side rails, to one side rail, or to other structures.

Ramps and Bridging Devices

A ramp is an inclined surface between different elevations for the passage of employees, vehicles, or both. A bridging device is a surface which spans a gap between a loading dock and a vehicle or between vehicles. It may be fixed, portable, adjustable, powered, or unpowered. It may also be called a car plate or dockboard.

Slip-resistant Floors

Slip-resistant flooring material such as textured, serrated, or punched surfaces and steel grating may increase slip-resistance. These types of floors are used in work areas that are generally slippery because of wet, oily, or dirty operations. Slip-resistant footwear may also be useful in reducing slipping hazards.

Effective housekeeping can minimize fall hazards where slippery surfaces are due to temporary or intermittent conditions. Use absorbents to clean up a spill where oily materials or corrosive liquids are accidentally spilled.

Personal Fall Protection

Once you know what measures your employer takes to prevent falls, find out what situations exist where the risk of falling can't be eliminated. Then you can begin to learn about personal fall protection systems designed specifically for each situation.

Personal Fall Arrest System

A personal fall arrest system is used to stop an employee safely after a fall from a working level. It consists of an anchor, connectors, a body belt or body harness and may include a lanyard, deceleration device, lifeline, or some combination of these.

Anchor: A secure point of attachment for lifelines, lanyards, or deceleration devices, that is independent of the means of support or suspension of an employee. That is, it is a separate point of attachment from any employee support lifeline attachment point.

The strength of any fall protection system is based on it being connected to a secure attachment point. When falling six feet, a person will exert up to 10 times their body weight as a shock load on the fall protection system. Therefore, attachment points must be capable of supporting at least 5,000 pounds per employee attached to the line or be designed as part of a complete engineered fall arrest system.

Connector: A device used to connect parts of the system together. It may be an independent component of the system, or it may be integral to part of the system (such as a buckle or dee-ring sewn into a body harness, or a snaphook spliced or sewn into a lanyard).

Harness: An arrangement of straps fastened such that the torso is supported during a fall. The attachment ring must be in the back of the harness near the shoulders. The anchor point of the lanyard or deceleration device should, if possible, be located above the wearer's harness attachment point. Hardware, except rivets, must be capable of withstanding a load of 5,000 pounds without cracking, breaking, or taking a permanent deformation.

Note: OSHA recommends that workers use full body harnesses instead of body belts. When subjected to an actual drop, a body harness distributes the shock wave more evenly over the body than does a belt. Similarly, the agency cautions employers against using non-locking type snaphooks.

Lanyard: A flexible line of rope, wire rope, or strap which generally has a connector at each end for connecting the body belt or harness to a deceleration device, lifeline, or anchorage. The lanyard may be a rope or shock-absorbing, or web, lanyard. It must be no longer than six feet. The shock-absorbing lanyard will substantially reduce the force created from arresting a fall.

Another type of lanyard is a self-retractable lanyard that allows freedom of movement but protects the worker should a fall occur. The webbing moves with the worker, pulling out when the person moves forward and retracting when the worker moves back. If the worker falls, the unit locks, restricting the fall distance to two feet or less.

Deceleration device: Any mechanism such as rope grabs, ripstitch lanyards, specially-woven lanyards, tearing or deforming lanyards, automatic self-retracting lifelines/lanyards, and so on, which serve to dissipate a substantial amount of energy during a fall arrest or otherwise limit the energy imposed on an employee during a fall arrest.

Lifeline: The lifeline may be either vertical or horizontal. Vertical lifelines may only support one worker at a time. Horizontal lifelines are subject to greater loads than vertical lifelines and must be properly designed and installed.

Self-retracting lifelines provide mobility as well as worker protection. The line retracts as the worker moves toward the unit and pulls out as the worker moves away. If the worker slips or falls, the sudden jerk on the cable activates the breaking mechanism and the worker is brought to a stop within two feet.

Note: When using a manufacturer's fall protection components or complete system, you must follow the manufacturer's recommendations and installation instructions. Your employer must train you on the safe use of the system. All systems must be inspected before use when installed, and before use each day. Systems must also be inspected at intervals as established by the manufacturer. In addition, your employer must set up a rescue program before you use any fall arrest system or use a system designed for self-rescue.

Positioning Device System

A positioning device system is a body belt or body harness system rigged to allow an employee to be supported on a wall, window sill, or other vertical surface and work with both hands free. A positioning device system supports the worker and is rigged so that the worker would not fall more than two feet.

Personal Fall Protection System for Climbing Activities

Personal fall protection systems for climbing activities are designed to keep climbers from being injured should they fall while ascending or descending. This equipment includes: ladder safety devices, limited velocity descent devices, automatic payout and self-retractive lifelines, and associated components.

Select the Right Equipment

You need to know what kind of equipment should be used for a given fall hazard. To determine this, you must learn:

- Where and when to use specific equipment and selection criteria.
- How to determine free-fall distance and total fall distance.
- Consideration of environmental and other workplace factors.

Careless or improper use of equipment can result in serious injury or death. Both you and your employer should know how to use personal fall arrest systems and follow manufacturer recommendations.

Use Personal Fall Protection Correctly

Before you use equipment and after any component change, learn to use the system safely. Find out:

- How to inspect equipment for mildew, wear, damage, and other deterioration before each use.
- Application limits, including how to estimate and limit the maximum arresting force to acceptable limits for the system.
- Methods of use, including intended functions and performance of equipment.
- How to put on, adjust, and connect the equipment.

- Anchoring and tie-off techniques.
- Emergency rescue plans and implementation.
- Maintenance procedures.
- Storage techniques.

Certain tie-offs (e.g., using knots, tying around sharp edges, etc.) can reduce system strength. This factor and maximum permitted free fall distance should be considered when determining the effectiveness of any personal fall protection system. Inspections before use, equipment limitations, and unique conditions at the worksite are also important.

Be able to recognize and avoid fall hazards you may encounter on the job. You should also be aware of general written policy/procedures on fall protection.

Supplier's Instructions

Your instructor will get comprehensive directions for the system's proper use and application from the supplier. Use this valuable information routinely. It will include things like:

- The force measured during the sample force test.
- Maximum elongation measured for lanyards during the force test.
- The deceleration distance measured for deceleration devices during the force test.
- Caution statements on critical use limitations.
- Application limits.
- Proper hook-ups, anchoring, and tie-off techniques, including the proper dee-ring or other attachment points to use.
- Proper climbing techniques.
- Methods of inspection, use, cleaning, and storage.
- Specific lifelines to be used.

Reporting Hazards

Reporting fall hazards is integral to any effective safety effort. Report unsafe equipment, conditions, or procedures. Equipment repair should receive top priority. Under no circumstances should defective fall protection equipment be used.

Work at Working Safely

Follow these guidelines when you have a risk of falling on the job:

- Make sure you have a guardrail or cover for all open pits, tanks, vats, and ditches.
- Use guardrails on all walks, runways, or platforms 4 feet or more from ground level, except on loading or unloading sides of platforms.
- Make sure you have a guardrail or cover for all floor openings and holes.
- Use personal fall arrest systems whenever you must work on powered platforms.
- Make sure guardrails, toeboards, and metallic mesh (or similar material) is on powered platforms used for exterior building maintenance.
- Use guardrails, gates, or mazes at all entrances and exits at floor landings affording access to man lift.

Foot Protection: Keeping on Your Toes

Understanding Foot Power

Support and propulsion are the two functions of the foot. Our feet permit us to walk, stand, sit, and kneel. They bear our weight when we jump, run, or reach above our heads.

The 26 bones in the foot are shaped in the form of an arch to provide a broad, strong support for the weight of the body. Because of how valuable our feet obviously are to us, we want to protect them from the hazards of the workplace.

Where Are the Regulations?

The Occupational Safety and Health Administration (OSHA) has developed regulations that specify foot protection to keep your feet safe at work. These regulations are located in 29 CFR

1910.132 and .136. OSHA requires that protective footwear meets the requirements of the American National Standards Institute (ANSI) consensus standard (ANSI Z41-1991).

Your employer must train you in the proper use of foot protection. You must know:

- When protective footwear is necessary.
- What footwear is necessary.
- How to properly put on, take off, adjust, and wear protective footwear.
- Protective footwear limitations.
- Proper care, maintenance, useful life, and disposal of protective footwear.

You must show you understand the training, and can use safety shoes properly before you will be allowed to perform work requiring their use.

A Real Pain in the Foot

Every day hundreds of workers in the United States suffer disabling injuries to their feet and toes. Foot and toe injuries numbered 91,800 in 1996, according to a 1997 Bureau of Labor Statistics report. This number represents about five percent of all disabling injuries. The foot is especially vulnerable to injury. For example, it's possible to severely sprain your ankle simply stepping off a curb! Yet many workers ignore the serious hazards in the workplace and refuse to wear protective footwear.

Some Types of Foot Injuries

Your feet are subject to many types of skin diseases, cuts, punctures, burns, sprains, and fractures. But sharp or heavy objects falling on the foot are the primary source of injury.

Other hazards include:

- **Compression** – the foot or toe is squeezed between two objects or rolled over.

- **Puncture** – a sharp object like a nail breaks through the sole.

- **Electricity** – a hazard in jobs where workers use power tools or electric equipment.

- **Slipping** – contact with surface hazards like oil, water, or chemicals causes falls.

- **Chemicals** – chemicals and solvents corrode ordinary safety shoes and can harm your feet.

- **Extreme heat or cold** – insulation or ventilation is required depending on climate.

- **Wetness** – the primary hazard may be slipping, but other hazards are discomfort and even fungal infections if your feet are wet for long periods of time.

Many plant operations or manufacturing processes involve a combination of hazards listed above. Yet, one study of workers who suffered foot injuries showed that less than 25 percent were wearing safety shoes or boots at the time of the accident. Many workers said that safety shoes were not normally used in their work and they felt foot protection wasn't needed.

Foot Protection Is Important

Foot protection is guarding your toes, ankles, and feet from injury. Manufacturers now offer a wide variety of protective devices for hazards in many industries. Manufacturers also continually update materials and engineering of their products to ensure protection from new hazards.

What About Safety Shoes?

What features make safety shoes different from regular "street" shoes? Basically, they are designed to protect the foot and toes in areas most likely to be injured. There is a safety shoe that offers protection from falling objects or weight pressing on the toes. A cushion between the toe cap and the foot offers comfort and insulation. A steel insole will protect the wearer from puncture wounds. Soles are made from a wide variety of materials, depending on the workplace hazards to be encountered by the person wearing them.

Some safety shoes have instep protection made of aluminum, steel, or plastic to protect the top of the foot and the front of the ankle. The safety shoe or boot can be insulated to protect from heat or cold. It may be waterproof or chemical resistant. Safety shoes are also available that offer ankle protection. With all these options specifically designed to protect your feet, can you honestly say that a pair of "street" shoes will do the same job of protecting your feet?

Some Specific Types of Safety Shoes

Safety shoes come in many varieties to suit very specific industrial applications. Here are descriptions of some types of safety footwear.

Safety Shoes

Standard safety shoes have toes that meet testing requirements found in the ANSI standard. Steel, reinforced plastic, and hard rubber are used for safety toes, depending on their intended use. These shoes are worn by workers in many types of general industry.

Metatarsal Guards

Shoes with metatarsal or instep guards protect the upper foot from impacts. In these shoes, metal guards extend over the foot rather than just over the toes.

Conductive Shoes

Conductive shoes permit the static electricity that builds up in the body of the wearer to drain off harmlessly into the ground.

By preventing accumulation of static electricity, most conductive shoes keep electrostatic discharge from igniting sensitive explosive mixtures. These shoes are often worn by workers in munitions facilities or refineries. Do not use these shoes if you work near open electrical circuits.

Safety Boots

Rubber or plastic safety boots offer protection against oil, water, acids, corrosives, and other industrial chemicals. They are also available with features like steel-toe caps, puncture-resistant insoles, and metatarsal guards. Some rubber boots are made to be pulled over regular safety shoes.

Electrical Hazard Shoes

Electrical hazard shoes offer insulation from electrical shock hazards from contact with open circuits of 600 volts or less under dry conditions. These shoes are used in areas where employees work on live or potentially live electrical circuits. The toebox is insulated from the shoe so there is no exposed metal. These shoes are most effective when dry and in good repair.

Sole Puncture Resistant Footwear

Puncture-resistant soles in safety shoes protect against hazards of stepping on sharp objects that can penetrate standard shoe soles. They are used primarily in construction work.

Static Dissipative Shoes

Static dissipative footwear is designed to reduce accumulation of excess static electricity by conducting body charge to ground while maintaining a sufficiently high level of resistance to protect you from electrical shock due to live electrical circuits.

Foundry Shoes

Foundry shoes are used by welders and molders in foundries or steel mills where there is a hazard from hot splashes of molten metal or flying sparks. Instead of laces, they have elastic gores to hold the top of the shoe close to the ankle. They can then be removed quickly in case hot metal or sparks get inside.

Add-On Foot Protection

Metatarsal guards and shoe covers can be attached to shoes for greater protection from falling objects. Strap-on wooden-soled sandals can be used for protection against the underfoot hazards of oils, acids, hot water, caustics, or sharp objects.

Rubber spats protect feet and ankles against chemicals. Puncture-proof inserts made of steel can be slipped into shoes to protect against underfoot hazards. Strap-on cleats fastened to your shoes will provide greater traction.

Be Sure Safety Footwear Meets Standards

When purchasing and selecting safety footwear, it is important to look for shoes and boots that meet the ANSI requirements. OSHA regulations state that safety shoes should meet ANSI standards (ANSI Z41-1991).

These standards set forth requirements for safety shoes in the areas of impact, compression, conductive, and puncture resistance performance. You always want to match the footwear to the job and the hazards you will be encountering there.

Purchase only those that are proven effective. As with any piece of safety equipment, the best time to evaluate safety footwear is before you purchase it. Even though safety footwear isn't cheap, when you look at how much you wear safety shoes, the cost becomes more reasonable.

FOOT

Work at Working Safely

Safety shoes can prevent serious, even disabling, injuries at relatively low cost. We read earlier that most workers' injuries occurred because they were not wearing protective footwear. As a review, let's look at some of the excuses that keep workers from using safety footwear. In each case, we'll "stamp out" the excuse with the facts!

They're ugly! – Some people are willing to sacrifice safety for style. However, safety shoes are now available in fashionable styles ranging all the way from running shoes to western boots.

They're too expensive! – When the cost is spread out over the life of the shoe, the price of safety is only pennies a day. Some employers contribute towards the purchase of safety shoes or allow employees to purchase footwear directly from the manufacturer. Whether you purchase safety shoes yourself, or receive help from your employer, the protection of safety shoes is well worth the cost.

They're not comfortable! – New designs in steel toe caps have reduced shoe weight and shoes now have inner soles of foam latex that "breathe" and make them more comfortable. Generally, the safety features of these shoes are not noticed at all until they are called upon to protect the foot of a worker during an accident.

I work in heavy industry. – Nothing will protect my feet if they're crushed or rolled over! First, safety shoes will lessen the severity of many injuries caused by heavy equipment or products. Also, there are still hazards present from falling tools and component pieces. A safety shoe will also reduce the danger from slipping, falling, and puncture injuries.

They're too clumsy for climbing! – The key here is to find a shoe suitable for working on ladders or scaffolding. Anyone who works in those situations should wear a shoe with a defined heel and good traction.

Steel toe caps will cut off my toes if crushed! – Protective toe caps in a quality shoe are tested in accordance with ANSI guidelines for protection. Toe caps are also designed to give a "buffer zone" of space over the toes in case they are crushed.

I don't know where to buy them! – In addition to stores that specialize in the sale of safety shoes, it possible to purchase footwear without leaving the job site or plant. Shoemobiles, in-plant stores, and catalog order programs bring the shoes right to you on the job.

There's no way to tell if a shoe will really protect my feet! – Purchase only those shoes that meet ANSI criteria. These shoes will have passed tough tests against many hazards, including falling and rolling objects.

I'm a safe worker, I won't have an accident! – This is probably the most dangerous excuse offered for not wearing safety equipment. "Accident" means an unexpected or unintentional event. No one can know what another worker will do or when a piece of equipment will malfunction. Safety shoes provide protection that is easy to wear, yet highly effective.

"It can't happen to me" is a dangerous myth that has been proven wrong again and again. So don't take a chance with your two good feet. Obtain proper safety footwear and wear it at all times on the job.

Hand Protection: Let Your Fingers Do the Working

How would you answer the question, "What is the most used tool in industry?" Some people would name a commonly used hand tool like a hammer or screwdriver. Others might respond with a list of larger equipment such as lathes or power tools. But the correct answer is deceptively simple. The most used tool in almost any workplace is the human hand.

Think of almost any job in your plant, from sweeping the stockroom floor to operating a computer. Your hands and fingers are the tools you use every working day. Try writing without using your thumb. Try holding a hammer with only two fingers. Hand protection is important because our hands are exposed to so many hazards in the workplace.

Where Are the Regulations?

The Occupational Safety and Health Administration (OSHA) regulates personal protective equipment in general and hand protection at 29 CFR 1910.132 and .138. OSHA requires your employer to select and provide you with hand protection when you are exposed to hazards such as skin absorption of harmful substances, severe cuts or lacerations, severe abrasions, punctures, chemical burns, or harmful temperature extremes. Currently, no consensus standard on hand protection exists.

Your employer must train you in the proper use of hand protection. You must know:

- When hand protection is necessary.
- What type is necessary.
- How to properly put on, take off, adjust, and wear gloves, mitts, or other protection.
- Hand protection limitations.
- Proper care, maintenance, useful life, and disposal of hand protection.

You must show that you understand the training, and can use hand protection properly before you will be allowed to perform work requiring its use.

Hand Injuries Are Common

Bureau of Labor Statistics data indicate that 228,200 workers suffered injuries to hands and fingers in 1996. That is, about 12 percent of work-related injuries are to hands or fingers.

At work, your hands are exposed to three basic kinds of hazards:

- **Mechanical hazards**. These are present wherever machinery is used. Injuries resulting from machinery use might include cuts, punctures, abrasions, or crushing.
- **Environmental hazards**. Factors like extreme heat or cold, electricity and materials handling have the potential to injure your hands.
- **Irritating substances**. Skin conditions such as dermatitis can be caused by contact with chemicals and biological agents (bacteria, fungi, and viruses). Chemicals and toxic substances can also enter the blood stream through abrasions or cuts.

The First Defense

The first defense in the battle to reduce hand injuries is engineering controls designed into equipment during manufacture or used to alter the work environment to make it safe and hazard free. Machine guards protect hands and fingers from moving parts and should not be altered or removed. Work stations and jobs should be designed to incorporate proper positions for tools, hands, and work objects.

Good Housekeeping Prevents Injury

Good housekeeping practices and personal cleanliness are also an important part of a preventive plan for hand protection. Wash stations and skin cleansers should be used along with other personal protection methods. Germicidal and antiseptic soaps, detergents and cleansing creams remove dirt, grease, oil, and chemicals from the skin after exposure.

Good housekeeping also applies to tools, plant equipment, and work areas. Cluttered aisles, poorly maintained machinery, and sloppy work habits can all contribute to accidents which can result in hand injury.

Types of Protective Equipment

Personal protective equipment (PPE) can help reduce the frequency and severity of hand and finger injury. Although fingers are harder to protect, they can be shielded from many common injuries. Personal protection is available in the form of gloves, mitts, finger cots, thimbles, hand pads, sleeves and hand lotions or barrier creams.

Gloves

Gloves are perhaps the most commonly used type of PPE. They provide protection to fingers, hands, and sometimes wrists and forearms. Ideally, gloves should be designed to protect against specific hazards of a job being performed. Types range from common canvas work gloves to highly specialized gloves used in specific industries.

Good examples of job-rated hand protection are the items designed for those who work with electricity – special rubber gloves and lineworkers' rubber insulating sleeves. The gloves are made of natural or synthetic rubber and are color coded to correspond with their level of voltage protection.

Rubber, vinyl, or neoprene gloves are also used when handling caustic chemicals like acids, cleansers or petroleum products. Leather gloves or leather reinforced with metal stitching are useful for handling rough or abrasive materials. Metal mesh gloves are worn by workers in the meat packing industry who work with sharp knives and saws.

Many gloves are rated as being safe for use with certain kinds of chemicals. If you are allowed to select your own PPE, read the glove manufacturer's chemical resistance charts. They rate each glove material and how it withstands specific chemicals.

Proper Fit Important

Wear only gloves that fit your hand. Gloves that are too small can tire your hands and gloves that are too large are clumsy to work with. Gloves should be worn with great caution near moving equipment or machinery parts. The glove could get caught and pull your fingers or hand into the machinery. Gloves should be given proper care and cleaning. They should be inspected regularly for change in shape, hardening, stretching, or rips.

Other PPE for the Hand

There are many other types of hand protection:

- **Mitts** are similar to gloves, but have a division for the thumb and one for fingers.
- **Finger cots** provide protection for a single finger or fingertip.
- **Thimbles** are used to protect the thumb or the thumb and first two fingers.
- **Hand pads** are used to protect the palm from cuts and friction. These pads are also used to protect against burns caused by hot objects. Heavier and less flexible than gloves or mitts, they should not be used for jobs requiring manual dexterity.
- **Sleeves or forearm cuffs** protect the wrists and arms. They are used to protect against heat, splashing liquids, impacts, or cuts.

Barrier Creams

Barrier creams or lotions can be used by themselves or along with other types of PPE. You might use a lotion when other types of protection cannot be used, such as when working with or near moving machinery. Three types of cream are available:

- **Vanishing cream** usually contains soap and emollients to coat the skin. They make cleanup easy and protect against mild acids.
- **Water-repellent cream** leaves an insoluble film on the skin. It protects against irritants in water – alkalis and acids.
- **Solvent-repellent cream** protects against irritating solvents and oils.

To be effective, creams or lotions should be applied frequently. Remember that these creams do not protect against highly corrosive substances.

A Look at Carpal Tunnel Syndrome

As discussed in the ergonomics chapter, Carpal Tunnel Syndrome (CTS) is a nerve problem of the hand and wrist. Repeated, forceful hand and wrist movements of some kinds can lead to pressure on the main nerve to the hand. Early symptoms of CTS are numbness and tingling in the fingertips.

Short, frequent breaks from repetitive tasks reduce the risk. A break for rest and exercise of the wrist, elbows and shoulders will increase circulation and allow the body to recover from repetitive movements. Keep the wrist in a straight position whenever possible, and reduce the speed and force of movements involving the wrist.

Minor cases of CTS may be cured simply with a few days rest, but serious, chronic cases may require surgery. However, surgery is expensive and does not guarantee long-term relief. Other CTS treatments include medication or splints which allow the worker to use the hand but prevent him or her from bending it.

Avoid wearing watches, bracelets or tight clothing that hampers wrist circulation. Grasp objects with the whole hand if possible. See a doctor immediately if CTS is diagnosed or suspected. Look for tools designed to ease strain on the hand and wrists.

What Happens if You Are Injured?

If you do get injured on the job, you should know what to do. For cuts, control the bleeding with direct pressure on the wound. For broken bones, immobilize the injured hand. For chemical or heat burns, put the hand under running water and flush for 10 to 20 minutes.

You might be faced with handling a more serious hand or finger injury like an amputation. Because severed limbs can often be reattached, act quickly. Control heavy bleeding or shock first. Keep the severed part cool, but don't freeze it. Make sure the injured person gets medical attention quickly. Do not apply a tourniquet unless the person is in danger of bleeding to death.

No matter what kind of injury occurs, get medical help as soon as possible. Report the accident to your supervisor and take the victim to your facility doctor, first-aid station, or hospital emergency room.

Work at Working Safely

Because you use your hands every day on the job, they can easily be injured. Keep these points in mind to protect your hands as you work:

1. The type of protective equipment you select will depend upon the nature of the hazard in your workplace.
2. Gloves should fit you properly and be maintained in the same careful way as other safety equipment.
3. Lotions or barrier creams may be used when other forms of protection are not practical. They should be reapplied frequently.
4. Lotions and barrier creams do not protect against highly corrosive materials.
5. Know the symptoms of carpal tunnel syndrome and seek medical attention immediately if you suspect this condition.
6. In the event of a hand injury, know proper first-aid procedures. Offer only the help you are trained to provide.

There are few activities on or off the job that don't involve your hands in some way. Driving, eating, writing, holding a loved one – the list could go on and on. Make sure your hands last a lifetime. Protect your most valuable tools – your hands.

HAND

Head Protection

Head injuries numbered 124,000 in 1996, according to a 1997 Bureau of Labor Statistics report. Injuries range from minor abrasions to death, and can include concussions, lacerations, trauma, burns, or even electrocution.

Head Protection Regulations

Head protection regulations for general industry are found in 29 CFR 1910.135. The standards recognized by OSHA for the design, construction, and use of protective headwear are in the American National Standard Institute's standard, ANSI Z89.1. *Safety Requirements for Industrial Head Protection*, ANSI Z89.1-1969, must be followed for helmets purchased before July 5, 1994. And, *Protective Headwear for Industrial Workers—Requirements*, ANSI Z89.1-1986, must be followed for helmets purchased after July 5, 1994.

How do Hard Hats Help?

Head injuries are caused by falling or flying objects or by bumping your head against a fixed object. Other causes are from electrical shock and burns. Hard hats are designed to do two things: resist the penetration and absorb the shock from a blow, and to provide protection from electrical shock and burn.

Hard hats lessen injury because the hard outer shell and the inner suspension system work together to absorb impacts. Hard hats that protect against electrical hazards are constructed of electrically insulating materials.

When to Use Hard Hats

When you are working in an area where there is a possible danger of head injury from impact from falling or flying objects, or where there is a risk of electrical shock and burns, you must wear your hard hat.

Types of PPE for the Head

Hard hats fall into two types and three classes. The classes are intended to provide protection against a specific hazardous condition.

The types include:

- Type 1—full brim.
- Type 2—no brim, but may include a peak extending forward over the eyes.

The classes of hard hats are:

- Class A—limited voltage protection.
- Class B—high-voltage protection.
- Class C—no voltage protection.

Class A

These hard hats are used for protection against the force of impact and penetration by falling objects. They also reduce the danger from low-voltage exposures.

Class B

This type of hard hat protects your head from the force of impact and penetration by falling objects. It also offers protection from exposure to high-voltage.

Class C

The design of these hard hats offers protection from the force of impact and penetration by falling objects. They are used where there is no danger from electrical hazards.

Each approved hard hat is marked on the inside of the shell with the manufacturer's name, the applicable ANSI designation, and the Class.

Some helmets are designed so they accept attachments such as face shields, hearing protection, winter liners, or lamps. If accessories are not used properly, the hard hat may not provide as much protection. Always follow the manufacturer's instructions.

Work at Working Safely

You should take proper care of your hard hat to prolong its life and your safety. The safety provided by a hard hat is limited if it is damaged or if it is not used as it was intended to be.

- Check you hard hat daily for signs of dents, cracks, or penetration. Do not use the helmet if any of these signs of damage are found. This inspection should include the shell, suspension, headband, sweatband, and any accessories.

- Do not store or carry your hard hat on the rear-window shelf of a car. Sunlight and high heat can degrade the helmet. Long periods of exposure to sunlight can lead to damage from ultraviolet rays. Signs of damage include a dulling, chalking, crazing, or flaking on the surface of the shell.

- Clean your hard hat once a month in warm, soapy water. Scrub and rinse the shell with clear, hot water (about 140 degrees F). Inspect the shell for damage after it has been cleaned.

- Do not paint your hard hat. Some types of paint and thinners may damage the shell or weaken the helmet. Consult with the hard hat's manufacturer for recommendations on using any type of solvent to clean paint, tars, oils, or other materials from the helmet.

Hearing Conservation: Now Hear This

In the past, workers accepted partial hearing loss as a cost of working in a noisy plant or on a production line. New workers were told by old-timers that they would soon get used to the noise. But times have changed. Noise is now recognized as an occupational hazard that can cause temporary or permanent hearing loss, stress, and other physical problems.

Where Are the Regulations

Regulations governing the allowable levels of noise and hearing protection requirements for employers to provide employees have been issued by the Occupational Safety and Health Administration (OSHA). The regulations can be found at 29 CFR 1910.95.

What Is Noise?

One definition describes noise as unwanted or unpleasant sound. We're all exposed to noise every day at home, at work, and in traffic. However, a clear definition of noise is hard to pin down. How you're affected by sound depends on several things—loudness and frequency of the sound, length of exposure, and even your age and health.

A temporary hearing loss can occur from a short exposure to loud sounds, but your hearing soon recovers when the noise stops. Despite this factor, one fact about noise is very clear. If the level is high enough for long enough, it can cause health problems—including permanent hearing loss.

How do we hear sounds? A sound source sends out vibrations into the air. These vibrations are called sound waves. The ear changes the energy in sound waves into nerve impulses which travel to the brain and are then interpreted.

Sound is measured by its frequency and intensity. Frequency is the pitch (high or low) of a sound. High-frequency sound can be more damaging to your hearing than low-frequency sound. Intensity is the loudness of a sound. Loudness is measured in decibels(dB), named for Alexander Graham Bell.

Intensity that exceeds an average of 85 dB over an eight-hour day may cause hearing loss. According to OSHA standards, workers may not be exposed to more than an average of 85 dB over an eight-hour period without hearing protection being provided.

Different Kinds of Noise

But noise is just noise, right? And too much of it is bad for your ears. It would be nice if the picture were that simple, but it's not. In general, there are three types of noise:

- Wide band is noise that is distributed over a wide range of frequencies. Examples are the noise produced in most manufacturing settings and by the operation of most internal combustion engines.

- Narrow band noise is restricted to a narrow range of frequencies. Examples include noise from various kinds of power tools, circular saws, fans and planers.
- Impulse noise is composed of temporary "beats" that can occur in on-and-off repeating patterns. Jack hammers, or power or punch presses are good examples of tools that cause impulse noise.

How Does Noise Hurt?

Hearing loss due to noise in the workplace is often underrated as an occupational hazard. It's frequently ignored because hearing loss usually takes place over a long period of time and may not be readily apparent until the damage is done. There's no visible wound or other sign of injury.

But too much noise can cause a variety of problems. It can make you tired and irritable from the strain of talking or trying to listen over loud sounds. You might not be able to hear important work or safety instructions because of excessive noise.

Evidence exists that other physical damage may occur because of the way the body reacts to noise. Let's say you're sitting in the facility's break area, absorbed in reading the sports page. The plant joker sneaks up behind you and – "BANG" –

pops a paper bag right behind your head. You're startled. Your muscles tense. Your heart beats rapidly and adrenaline starts to flow. Your body prepares for "fight or flight," a natural response to a sudden noise.

Once you realize that there's no danger, you settle down. Your body slowly returns to normal. Now if this would happen over and over, every day, your body would suffer. This is why studies link noise with high blood pressure, ulcers, headaches, and sleeping disorders. Add these potential dangers to the obvious damage that noise can do in causing either temporary or permanent hearing loss.

Don't you believe it if you're told that you'll "get used" to all the noise in your workplace. High levels of noise are dangerous, and the situation needs to be looked at carefully if you find yourself in this kind of a work environment.

Noise Can Be Controlled

If the noise level in your work area is too high, your employer must take additional action to reduce that noise. Steps to reduce noise might include:

- Moving noisy machinery to a separate area away from as many workers as possible, or building a sound barrier around it.

- Placing machinery on rubber mountings to reduce vibration.

- Using sound-absorbing acoustical tiles and blankets on floors, walls and ceilings.

- Arranging work schedules to cut down on the time each worker spends in a noisy area.

Manufacturers have also responded to requests to meet noise specifications at the installation and operation level. Since a variety of machinery and equipment can add noise to the workplace, workers, manufacturers and plant operators must all cooperate to reduce noise levels in the workplace.

Some equipment like saws and punch presses just can't be made to run any quieter, so it's up to you to protect yourself with the proper hearing protection if you work with that machinery. Don't take a chance with your hearing.

However, if workers are still exposed to hazardous levels of noise after such controls have been put in place, employers must provide hearing protective devices.

Protect Your Hearing

Various kinds of hearing protective devices are available for use in the workplace. The selection of the right hearing protection depends on several factors:

The noise hazard – what noise levels will you be dealing with?

Frequency of the noise – will it be continuous or intermittent? (Some earplugs or muffs reduce the force of noise (attenuate) better at lower frequencies than at the higher frequencies.)

Fit and comfort – the protective devices must fit properly and be comfortable enough to wear as long as they are needed.

Noise Reduction Rating or NRR – all hearing protectors carry a label indicating the NRR; a higher number on the label means more effectiveness.

Types of Hearing Protection

Your employer will conduct a hazard assessment to determine what types of noise control measures are needed. The first lines of personal defense against excessive noise are engineering and administrative controls. After taking these measures, your employer will determine what types of hearing protection devices (HPDs) will complete your protection from your facility's specific noise hazards.

Hearing protectors filter out the loud noise. This means they do not block out sound completely, but they reduce the amount of sound reaching the delicate parts of the ear. By doing so, they offer some protection.

HEARING

With protection, your hearing will not get overloaded by the surrounding noises (glare) that interfere with speech and machinery sounds. Four categories of HPDs are available:

Enclosure

The enclosure type hearing protection completely surrounds the head like an astronaut's helmet. This type of protection is not too popular due to its cost and the discomfort factor caused by the size and weight of the helmet.

Earplugs

Earplugs, also known as aurals, fit in the ear canal. They come in three forms:

- **Custom-molded** earplugs are made for specific individuals, molded to the exact shape of that person's ear. Silicone rubber or plastic molding compound is placed in each ear and allowed to set; these may then be used directly as earplugs or serve as molds for the final plugs.

- **Molded inserts** often called pre-molded, made from soft silicone rubber or plastic, are reusable and should be kept very clean to avoid infection. Use warm, soapy water to clean them after each use, and store them in a carrying case.

- **Formable** plugs fit all ears. Made of foam rubber, waxed cotton, or acoustical fibers, they are disposable.

Canal Caps

Canal caps (also known as superaural) seal the external edge of the ear canal to reduce sound. The caps are made of a soft, rubber-like substance and are held in place by a headband. This type of ear protection is a good alternative for those who can't use earplugs or for workers who enter and leave high noise areas frequently during the course of their work day.

Earmuffs

Earmuffs (also known as circumaural) fit over the whole ear to seal out noise. Earmuffs usually reduce sound levels by 20 to 25 dB. A typical muff is made up of three basic parts—cups, cushions, and headband. The cups are made of molded plastic and are filled with foam or other material. They vary in size and are adjustable. The cushions are covered with plastic and filled with liquid, air, or foam. Liquid or grease-filled cushions give better noise protection than plastic or foam types, but can be prone to leakage. The headband simply holds the cups against the head. It may be worn over the head, behind the neck or under the chin.

There are also specialty earmuffs for different job requirements. Dielectric muffs have no metal parts for those workers exposed to high voltages. Electronic earmuffs reduce hazardous noise but magnify wanted sounds like voices. Folding earmuffs are designed for use in situations where protection isn't required full-time but must be quickly available when needed. Cap-mounted muffs are attached directly to safety hats.

How Effective Are They?

In general, earplugs can reduce noise reaching the ear by 25 to 30 dB in the higher sound frequencies, generally considered to be the most harmful. Earmuffs can reduce noise 20 to 25 dB. Combinations of the two protectors can give 3 to 5 dB more protection. No matter what type of protection device you consider, remember that the only effective hearing protector is the one that you wear!

Audiometric Testing Is Vital

It is very important to keep track of your hearing by having it tested periodically. An audiometric test is a procedure for checking a person's hearing. Employers with facilities where noise exposure equals or exceeds an average of 85 dBA over an eight-hour day are required to provide their employees with audiometric testing.

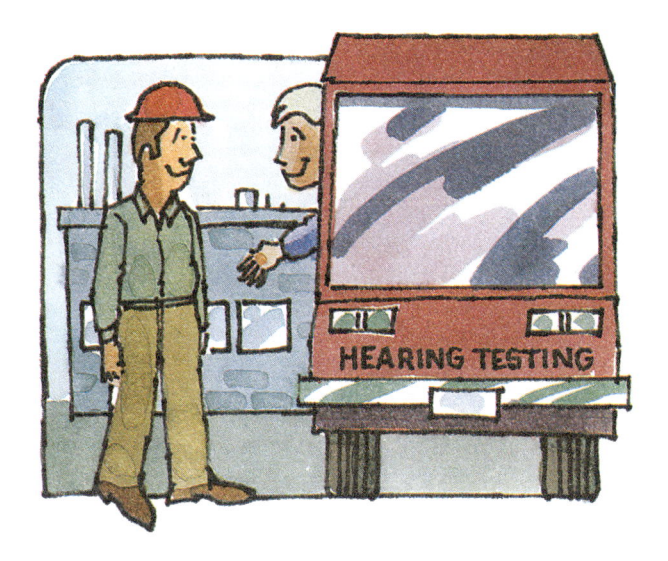

A trained technician uses an instrument (an audiometer) to send sounds (tones) through headphones. The person being tested responds to the test sounds when they are first heard. The chart that records responses to the test sounds is called an audiogram.

This test is an important part of your effort to conserve and protect your hearing. It checks hearing ability so that any hearing loss can be identified and dealt with properly and promptly. Have your hearing tested periodically when you're transferred to a noisy work environment or when you're exposed to noise levels that may be dangerous to your hearing.

It's Your Hearing – Keep It

Some people argue that hearing loss is a natural result of aging and the generally noisy environment in which we live. Noise is a real occupational hazard for many workers. It has been estimated that as many as 30 million workers are exposed to unwanted noise over an extended period of time. It's up to you to know the hazards and how to protect yourself. Because hearing loss is such a gradual process, it's easy to overlook or ignore the value of hearing protection.

Work at Working Safely

You are ultimately responsible for protecting your own hearing. You have the most to lose if you suffer hearing loss as a result of on-the-job noise hazards. Let's review a few important reminders about hearing conservation:

1. Disposable earplugs may be more convenient to use than long-term use plugs, but make sure they fit you correctly so that they will be effective.

2. Employees whose noise exposure equals or exceeds 85 dBA over an eight-hour period are required to have an annual audiometric test to check their hearing.

3. Keep hearing protectors in good operational order with routine maintenance and replacement of defective parts.

4. Don't use homemade hearing protectors such as wadded cotton or paper. They don't work.

5. Wear ear protection at home for any noisy job like operating a chainsaw or using various kinds of shop equipment. And watch the volume on your stereo headphones or portable radio/tape player headphones.

The sounds of everyday life—nature, music, the voices of family and friends all add pleasure and meaning to our lives. Value them enough to protect your hearing from damage or loss!

HEARING

Respiratory Protection: Choosing and Using Respirators

Respirators are important types of personal protective equipment (PPE). When ventilation or other engineering controls are not adequate to keep the air safe, a respirator will protect you.

Where Are the Regulations?

The Occupational Safety and Health Administration (OSHA) regulates the use of respirators in general industry, construction, and other industries. The regulation is at 29 CFR 1910.134. OSHA has several other regulations on certain chemicals or activities that require the use of respirators. (For example, the regulation on respiratory protection for exposure to Tuberculosis in the health care industry is at 29 CFR 1910.139.) This chapter focuses on the requirements of 29 CFR 1910.134.

Respiratory Protection Program

Your employer must have a worksite-specific respiratory protection program. The program is needed when respirators are required. The program includes:

- Procedures for selecting respirators.
- Medical evaluations.
- Fit testing for tight-fitting respirators.
- Procedures for proper use of respirators.
- Schedules and procedures for cleaning, disinfecting, storing, inspecting, repairing, discarding, and otherwise maintaining respirators.
- Procedures on safe air quality for atmosphere-supplying respirators.
- Training in respiratory hazards.
- Training in respirator use, maintenance, and the limitations of respirators.
- Procedures to make sure the program is effective.

If your employer allows you to wear a respirator on a volunteer basis (when the contaminants are at safe levels), the program only includes provisions for medical evaluations, cleaning, storage, and maintenance. (No program is needed if only dust masks are used on a volunteer basis.)

Who Needs to Wear a Respirator?

You can't always see, smell, or taste the dust, smoke, mist, fumes, sprays, vapors, or gases that can be hazardous to your health. Your employer is responsible for determining when respirators are needed in the workplace. That decision is based on:

- The exposure limits of the contaminants—safe levels that employees can be exposed to each workday without risking health problems.
- Scientific measurement of the exposure levels in the work area.
- The amount of oxygen in the work area.

What Types of Respirators Are Available?

There are two respirator types: air-purifying and atmosphere-supplying.

Air-purifying

These respirators remove the contaminants from the air as you breath. There must be enough oxygen in the air when using an air-purifying respirator. Typically, the respirator has a tight-fitting facepiece. As you inhale, a "negative pressure" suction inside of the facepiece forces the outside air through the respirator's cleaning elements before you breath it. You must use a filter, cartridge, or canister that is approved for the dust, mist, fume, aerosol, chemical vapor, or gas that you are exposed to.

Air-purifying respirators have limits on how long they will protect you. Filter media and sorbents get clogged or saturated as they are used. Follow your facility's cartridge or filter change schedule. The filters or cartridges can be used up before you notice a smell or taste inside of the respirator.

Warning: Not all types of hazardous substances can be safely removed by air-purifying respirators. Sometimes the amount of the contaminant in the air is too high for the filters or cartridges to be effective. Sometimes the contaminants are unknown. In some situations, there is not enough oxygen in the work space. In these cases, you'll need an atmosphere-supplying respirator.

Atmosphere-supplying

Atmosphere-supplying respirators provide you with breathing air from a clean source. Supplied-Air Respirators (SARs) use an airline to feed clean air to the respirator's facepiece, helmet, or hood. A Self-Contained Breathing Apparatus (SCBA) supplies air from tanks that are carried by the user. Fire fighters typically wear SCBAs.

Atmosphere-supplying respirators blow air into a facepiece, helmet, or hood to create a "positive pressure" that prevents the hazardous air from leaking in. Frequently, these respirators are operated in the "pressure-demand" mode—this means that the air is regulated so more air is supplied as the wearer inhales.

People wearing SARs are limited in how much they can move around because of the airline. SCBAs are portable, but the tanks have limited operating times. Careful attention must be paid to the quality of the breathing air being supplied to the wearer of any atmosphere-supplying respirator.

RESPIRATORY

Are You Physically Able to Wear a Respirator?

Wearing a respirator adds a physical burden to the job. Before you can be fit tested or required to wear a respirator, you must be evaluated by a physician or other licensed health care professional (PLHCP). Your employer will tell the PLHCP about the type of respirator you will be wearing and your job conditions. You will need to fill out a medical questionnaire. You might need an exam and medical tests. The PLHCP will give you and your employer a written statement on your fitness to use the respirator. The PLHCP can limit how you use a respirator and can require periodic follow-up medical evaluations. Follow-up evaluations can also be done after any changes in your respiratory protection program.

Monitor your health after the initial medical evaluation. Report any medical signs or symptoms that would have an effect on your ability to wear a respirator. Some examples are: shortness of breath, dizziness, chest pain, chest injuries, lung diseases, or heart conditions. These conditions require a follow-up medical evaluation to determine if you can continue to use a respirator.

Selecting a Respirator

Selecting the right respirator for the job is an important decision. Your employer has to match the capabilities and limitations of the respirator to the hazards of the job. All respirators must be certified by the National Institute for Occupational Safety and Health (NIOSH), and must be used as they were intended to be used.

IDLH Atmospheres

If the respirator is to be used in an Immediately Dangerous to Life or Health (IDLH) atmosphere, it must be either:

- A full facepiece pressure demand SCBA certified by NIOSH for a minimum service life of thirty minutes, or
- A combination full facepiece pressure demand supplied-air respirator (SAR) with auxiliary self-contained air supply.

Non-IDLH Atmospheres

If the respirator is to be used in non-IDLH atmospheres (where there is enough oxygen, and where contaminant levels are known) the respirator must be capable of protecting you from the contaminant levels that are expected during the job and in reasonably foreseeable emergency situations. Atmosphere-supplying respirators are appropriate for non-IDLH atmospheres.

RESPIRATORY

For protection against gases and vapors, the employer can provide an air-purifying respirator in some situations. The canisters and cartridges must be approved for the hazards. If the air-purifying respirator does not have and end-of-service-life indicator that automatically warns the wearer that the canister or cartridges need to be changed, the employer must have a strict schedule for changing the canisters or cartridges.

Air-purifying respirators can also be used to protect against particulates. NIOSH Type 100 certified filters (or NIOSH certified High Efficiency Particulate Air (HEPA) filters) offer protection from some of the most hazardous particulates. Other types of filters are approved for use with various types and sizes of particulates. NIOSH approval codes are based upon the filtering efficiency level and the filter's effectiveness on oil-based particulates.

How to Fit a Respirator

Before you can be required to use a respirator with a tight-fitting facepiece, you must pass a fit test. You cannot be fit tested if you have any facial hair growth (stubble beard growth, beard, mustache, or sideburns) between the facepiece sealing surface and your skin. The fit test has to be done using the same type and size of respirator that you will be wearing in the workplace. You must have the fit re-tested at least annually. If you've passed the fit test, but feel that the respirator doesn't fit later on, you can have another test. OSHA's procedures for conducting the fit test must be followed. The type of test depends on the type of respirator you are using.

Using Your Respirator

Whenever you are wearing a respirator, make sure that you always leave the respirator use area:

- if you detect vapor or gas breakthrough, changes in breathing resistance, or facepiece leakage.
- if you need to change filter, cartridge, or canister elements.
- if you need to wash your face or the facepiece to prevent eye or skin irritation.

- if your respirator needs repairs.

A good fit is important each time you wear a respirator with a tight-fitting facepiece. Your employer cannot allow you to wear a respirator with a tight-fitting facepiece if you have any condition, including facial hair, that interferes with the face to facepiece seal or the respirator's valve function. Glasses or goggles must be worn so that they do not interfere with the the seal. To make sure that your respirator has a good seal, you must perform two seal checks each time that you put on your respirator. General instructions for these checks are:

- Positive pressure check—Close off the exhalation valve and exhale gently into the facepiece. The seal is good if you feel a slight positive pressure built up inside the facepiece without air leaking out around the seal.

- Negative pressure check—Close off the inlet opening of the canister or cartridges by covering with your palm(s) or by replacing the filter seals. Inhale gently so that the facepiece collapses slightly, and hold your breath for ten seconds. The seal is good if the facepiece remains slightly collapsed and there is no air leaking in around the seal.

If employees will be entering areas with IDLH atmospheres, at least one employee who is trained and equipped to provide rescue must remain outside and maintain communication with the employee(s) in the IDLH area. In addition, during interior structural firefighting, at least two employees equipped with SCBA must enter the IDLH atmosphere together, and they must stay in contact with each other. Also, at least two people trained and equipped for rescue must remain outside during firefighting.

Know what to do in an emergency situation. If your respirator malfunctions while you are in a hazardous atmosphere, leave immediately. If your respirator is equipped with an auxiliary self-contained air supply, use it as you exit to safety. Some employees may be trained and authorized to respond in emergency situations that require the use of a respirator (for example, chemical releases, confined space rescues, or firefighting). These employees may respond within the scope of their specialized training—everyone else must follow emergency plans to evacuate.

Care and Maintenance of Respirators

Put a priority on keeping your respirator clean and in good repair. Don't risk irritations, disease, or contamination from using a dirty or damaged respirator. Keep it ready to use.

If you have been issued your own respirator, clean and disinfect it as often as necessary to keep it clean. If a respirator is shared, clean it before the next person uses it. If a respirator is kept for emergency use only, clean it after each use. When you're done with fit testing or training exercises, clean those respirators, too.

Store your respirator so it is protected from damage, contamination, dust, sunlight, extreme temperatures, excessive moisture, and damaging chemicals. Facepieces and other parts can be permanently deformed if they are smashed out of shape during storage. After it's deformed, your respirator won't fit you anymore.

Inspect your respirator before each use, and inspect it again when you clean it. SCBAs must be inspected monthly. Air cylinders must be kept fully charged. Respirators kept for emergency use must be inspected each month and are to be stored in clearly marked compartments or covers.

Check how the respirator is functioning. Look at the condition of the facepiece, head straps, valves, connecting tube, cartridges, canisters, filters, etc. Make sure rubber parts are pliable and are not cracking. Have repairs made by a qualified person using manufacturer-approved parts. Get repairs done before you need to wear the respirator again.

Respirator Users Need Training

Training helps you to understand why the respiratory protection program is set up the way it is and shows you how to get the most protection from the equipment. You must have training before any required respirator use. Retraining is required at least annually. Changes in the workplace can also trigger retraining. If you are using a respirator on a voluntary basis where contaminants are at a safe level, your employer has to provide you with basic information on respirator use.

Work at Working Safely

1. Only use respirators that have been selected to protect against the hazards.
2. When you put on a respirator, check its fit and operation.
3. Look for respirator damage or deterioration before and after use.
4. Respirators need to be cleaned, disinfected, and stored properly. Follow filter, cartridge, and canister change schedules.

SLIPS, TRIPS AND FALLS: ON THE JOB SAFETY BASICS

In this culture, falls are often not taken seriously. On TV and in cartoons, spectacular falls are done for special effects or to make us laugh. A person falls but doesn't get a scratch.

In reality, falls are accidents which often cause injury and lost time. Falls can even be fatal. Falls caused 684 (11 percent) workplace deaths in 1996 (Bureau of Labor Statistics, 1997), along with 317,900 (17 percent) injuries. In the same year, slips and trips caused 59,300 workplace injuries.

Falls are costly accidents. In the past, workers involved in falls have lost a median of 8.5 work days because of their accidents. Injuries from falls may include cuts, bruises, muscle sprains and strains, broken bones, and back injuries.

SLIPS, TRIPS

Physical Factors at Work in a Fall

It might seem that an accident due to a loss of balance is pretty uncomplicated. Actually, slips, trips, and falls involve three laws of science:

Friction is the resistance between things, such as between your shoes and the surface you walk on. Without it, you are likely to slip and fall. A good example is a slip on ice, where your shoes can't "grip" the surface, you lose traction and you fall.

Momentum is affected by speed and size of the moving object. You've heard the expression, "The bigger they are, the harder they fall." Translate that to mean the more you weigh and the faster you are moving, the harder your fall will be if you should trip or slip.

Gravity is the force that pulls you to the ground once a fall is in process. If you lose your balance and begin to fall, you're going to hit the ground. Your body has automatic systems for keeping its balance. Your eyes, ears, and muscles all work to keep your body close to its natural center of balance. A fall is likely if your center of balance (sometimes called center of gravity) shifts too far and can't be restored to normal.

What Happens When You Slip?

Slips are a loss of balance caused by too little friction between your feet and the surface you walk or work on. Loss of traction is the leading cause of workplace slips.

Slips can be caused by wet surfaces, spills, or weather hazards like ice and snow. Slips are more likely to occur when you hurry or run, wear the wrong kind of shoes, or don't pay attention to where you're walking. Follow these safety precautions in order to avoid a slip:

- Practice safe walking skills. If you must walk on wet surfaces, take short steps to keep your center of balance under you, and point your feet slightly outward. Move slowly and pay attention to the surface you're walking on.

- Clean up spills right away. Whenever you see any kind of spill, clean it up yourself or report it to a maintenance person. Even minor spills can be very hazardous.

- Don't let grease accumulate on a shop floor around machinery. If grease is present in your work area, be sure that it's cleaned up promptly.

SLIPS, TRIPS

- Be more cautious on smooth surfaces. Move slowly on floors which have been waxed but not buffed, and other very slippery surfaces.

Wearing the Right Shoes Helps

One of the best ways to help prevent slip, trip, and fall injuries is to increase friction between your shoes and the surfaces you walk on. The amount of traction a sole provides varies with the work surface.

For instance, shoes with neoprene soles can be used safely on most wet or dry work surfaces. However, they are not recommended for oily conditions.

When selecting safety shoes, you have to determine what conditions and/or hazards you face most often on the job. Non-slip shoes and soles will also be useful when climbing ladders or scaffolds. Be sure your footwear matches the working conditions present on your job.

Other devices are available to increase traction on your shoes. Strap-on cleats can be fastened to soles for greater traction on ice. There are non-skid sandals and boots that slip over shoes and offer better traction on ice, oil, chemicals, and grease. If these devices are available, use them as the job requires.

Properly Cleaned Floors Help

OSHA requires that the workplace be kept clean and orderly (29 CFR 1910.22(a)). Floors must be dry and free of protruding objects such as nails, splinters, holes, or loose boards.

Many slip accidents are caused by improper cleaning methods. The average person takes thousands of steps each day on a wide variety of floor surfaces. Floors can be treated with many finishes and you might have occasion to select products to be used in clean-up or maintenance.

A floor finish can be used to make an unsafe floor less slippery or can accidentally be used to make a safe floor dangerous. For instance, carnauba is a paste-wax product used to give floors a brilliant shine. When used properly, it is safe and effective. But when a build-up of this wax occurs on floors, it can produce a very slippery surface.

Acrylic finishes are available for problem floor surfaces like terrazzo and marble. One manufacturer makes an acrylic finish that contains carborundum flakes for greater traction. This finish is particularly useful when applied in restaurants and other locations where floors are often wet. Pine tar disinfectants used on ceramic floors sometimes leave a slippery residue and can cause bathroom falls.

Floors should be cleaned only with clean water. If soaps or commercial strippers are used on a floor, be sure that no residue remains when the floor dries. If you're mopping or cleaning, post signs or place barricades to warn others of a wet surface.

Added Traction for Wet Floors

One way to avoid slips on frequently wet surfaces is to apply some type of abrasive that will increase traction. Epoxies and enamels that contain gritty compounds may be painted on concrete, wood, and metal. These products are especially useful for aisles, walkways, ramps, and loading docks.

Some manufacturers offer a selection of strips and rolls of skid-resistant material that you can apply to stairs, gangplanks or other potentially hazardous walking surfaces. Rubber mats can be used as a permanent or temporary solution to slippery areas. Use them in kitchens, shower and locker rooms, or near building entrances where water or other materials can be carried in from outdoors.

What Happens When You Trip?

Trips occur whenever your foot hits an object and you are moving with enough momentum to be thrown off balance. A trip can happen when your work area is cluttered, when lighting is poor, or when an area has loose footing. Trips are more likely to happen when you are in a hurry and don't pay attention to where you're going. Remember these rules to avoid tripping:

- Make sure you can see where you're going. Carry only loads that you can see over.
- Keep work areas well-lit. Turned-off lights and burned-out bulbs can interfere with your ability to see clearly. Don't grope in the dark. Use a flashlight or extension light to make your walking area visible in unlighted areas.
- Keep your work area clean and don't clutter aisles or stairs. Store materials and tools in closets, cabinets, or specially assigned storage areas.
- Arrange furniture so that it doesn't interfere with walkways or pedestrian traffic in your area.
- Extension or power tool cords can be dangerous tripping hazards. Tape them to the floor or arrange them so that they won't be in the way for pedestrians.
- Eliminate hazards due to loose footing on stairs, steps, and floors. Report loose carpeting, stair treads, or hand rails. Broken pavement and floor boards or loose floor tiles can also catch a foot and cause a fall.
- On loading docks, store gangplanks and ramps properly.

Stairs Can Be Dangerous

Another high-risk area for the average worker is stairs. Loss of traction causes the highest number of stairway slipping and falling accidents and is usually due to water or other liquid on the steps. Because we use stairs so often, it's easy to forget that they can be hazardous. You can protect yourself from injury:

- Use handrails whenever possible. If you are carrying something and can't grip the rail, use extra caution.
- Don't run up or down stairs or jump from landing to landing.
- Don't carry a load that you can't see over.
- Report any unsafe conditions promptly. Maybe you can't control lighting or a cluttered stairway, but you can report them to your supervisor or maintenance staff.
- Report broken stair treads, floor boards, or handrails.

What Happens When You Fall?

Falls occur whenever you move too far off your center of balance. Slips and trips often push you off your center of balance far enough to cause a fall, but there are many other ways to fall. They are also caused by makeshift ladders, misuse of ladders, accidents while climbing, and improper scaffolding use.

Most falls are from slips or trips at ground level, but falls from greater heights pose a much higher risk of serious injury. Avoid falls of any kind with these safety measures:

- Don't jump. Lower yourself carefully from docks, trucks, or work stages.
- Check lighting. Make sure hallways, stairs, and work areas are properly lit.
- Repair or replace stairs or handrails that are loose or broken. If maintenance isn't your job, report these hazards to the proper personnel in your company.
- Don't store things on stairs or in aisles.
- Wear good shoes. Non-skid soles are a good choice. Remember that high heels or platforms are less stable than flat shoes.

A Few Words About Ladders

A ladder can be a great help on the job. While it is uncomplicated and simple to use, you shouldn't take ladder safety for granted. The following points summarize many of OSHA's regulations for ladders and can serve as guidelines for ladder use.

- Don't build makeshift ladders out of chairs, benches, or boxes. If the job calls for a ladder, take the time to find one.

- Make sure there's only one person on a ladder at a time.
- Check the ladder's condition before climbing. Don't use a ladder with broken or cracked rails or rungs. If the rungs are slippery with grease or oil, clean them. The ladder should have safety feet.
- Don't place a ladder on boxes or blocks to make it taller. Inspect all ladders for defects before you begin climbing. Face front and use both hands as you climb.

- Don't overreach from a ladder. If your waist reaches past the uprights, you've gone too far—move the ladder.

- Set ladders up properly by using the 4 to 1 rule. The distance from the wall to the base of the ladder should be one fourth the distance from the base of the ladder to where it touches the wall.

- Hoist tools or materials up to you after you reach the top of the ladder. You need both hands for climbing.

- Don't stand on top of a step ladder. Also be careful not to get too close to the top of an extension or straight ladder.

For more information on ladders, check the OSHA regulations which can be found at 29 CFR 1910.25 and 1910.26.

SLIPS, TRIPS

Scaffolds Can Be Dangerous

Scaffolds are elevated work platforms that may be built up from the ground, attached to ladders or suspended from above. If you use scaffolds in your daily work, you know that they demand respect. Check OSHA regulations at 29 CFR 1910.28 for detailed requirements on using scaffolds. A fall from scaffolding can result in serious injury or death. Keep both feet firmly on the scaffold with these safety precautions:

Make sure scaffolds are sturdy. Check them daily for any safety defects. Always clear work surfaces of snow, ice, or slippery materials. Spread sand onto wet planking for sure footing.

Never overload scaffolds with people, equipment, or supplies. Lock casters on mobile scaffolding to prevent movement when working. Use ladder jack scaffolds only for light-duty work. Fall and climbing protection devices prevent injury in the event of an accidental slip, trip, or fall on ladders or scaffolding. If your job requires you to wear such a fall protection device, know how it works and keep it in good working condition.

Scaffold accidents occur from improper use or poor maintenance. Respect scaffolds and they will keep you safe.

Work at Working Safely

Preventing slips, trips, and falls is a task that depends on many factors—most importantly—you. You might not be able to change your workplace, but you can recognize dangers, work to eliminate hazards, and use safety devices and equipment. Take these precautions as you work:

1. Use belts, hard hats, safety shoes, and hand rails for safety.
2. Watch where you're going and look for hazards in your way.
3. Move carefully on stairs, in hallways, aisles, and work areas.
4. Report hazards like poor lighting, spills, and broken stairs.
5. Learn how to set up and use ladders and scaffolding safely.

Remember that falls are a leading cause of injuries. They aren't funny—preventing them is serious business. Do your best to avoid slips, trips, and falls.

VIOLENCE IN THE WORKPLACE

Workplace violence can range from verbal or physical threats or intimidation to assault and battery. Nearly a million workers are assaulted in the workplace each year, resulting in an estimated 159,000 injuries. In addition, there were 912 workplace homicides in 1996, accounting for 15 percent of the total number of U.S. workplace fatalities that year. Homicide is the number one killer of women in the workplace, and the second leading cause of death for all American workers. The Occupational Safety and Health Administration (OSHA), the National Institute for Occupational Safety and Health (NIOSH), the United States Office of Personnel Management, and several states have issued guidelines and preventative strategies for the control of workplace violence. This chapter will suggest ways for you to prevent violence at your job and ways to respond to violence if it should occur.

Sources of Violence

Occasionally, violence occurs from within an organization. Stress from work assignments, performance reviews, changes in policies, etc. can lead to disputes among co-workers, supervisors, and management. When these issues go unresolved, arguments, threats, harassment, vandalism, arson, assault, or other violent acts can be the result.

The greatest risk of work-related homicide comes from violence inflicted by third parties, such as robbers and muggers. Robbery and other crimes were the motive in 80 percent of all workplace homicides in 1996. Incidents of domestic violence or violent acts by disgruntled clients, customers, patients, or former employees can also find a way into the workplace.

Risks have been identified with workplaces involved in dealing with the public; exchanging money or guarding valuables; delivering passengers, goods, or services; working out of a mobile worksite (taxicab or police cruiser); working with unstable or volatile persons in health care or social service settings; working late at night or early in the morning; working alone or in small numbers; or working in high crime areas.

Avoid Becoming a Target

Think ahead to identify your risks and plan to avoid them. Everyone in your organization can work together to prevent violence. Report all violent incidents so measures can be taken to keep them from being repeated.

Recognize Potentially Violent Situations

Learn to recognize situations that could result in violence. Often a co-worker, customer, or client will express troubled feelings before becoming angry or violent. Many times, listening and concern is all that is needed at the early stages of trouble. If you sense that the problem will get worse, or if someone threatens you, take it seriously and report the incident to your supervisor.

Be aware of places where an assailant could hide, and be leery of someone who is loitering or does not belong in the workplace.

Take Preventative Measures

When you have to work alone or in a small isolated group, use the "buddy system" or ask for an escort. Radios or cellular phones can help you to stay in touch. If you travel between work sites, call to check in so your co-workers know where you are, and keep your vehicle well maintained to avoid breakdowns. When walking to your vehicle, have your keys ready, stay in well lit areas, appear confident, and avoid "hiding spots" or threatening strangers. Look in and around your vehicle before getting in, and lock your doors before you start your vehicle. If you are followed or threatened along the way, keep going to the closest safe area with people and a phone.

Many workplaces have card-key access systems and have employees wear identification badges. Visitors have to check in to get a pass before they are allowed inside. Know who is, and who isn't, supposed to be in your work area. If you are working with the public, limit your customers or clients to using doorways that can be seen by workers.

Maintain good visibility in your workplace. Windows and mirrors help eliminate "hiding spots." Windows also help police or security guards to see in and check on your safety. Good lighting inside and out will deter criminals.

Video cameras, alarms, metal detectors, and security patrols reduce crime. Use drop safes and post signs to show that little cash is on-hand. Vary the schedules for money drops and pick-ups to make it harder on robbers. Keep valuable personal items locked up or at home. Counters and windows act as physical barriers between workers and potential attackers.

Handling a Violent Situation

If you become involved in a violent situation:

- Report threats or suspicious activity to your supervisors or the police. Keep your distance from the situation. Stay safe, and let the authorities handle it.

- If you are confronted, talk to the person. Stay calm, maintain eye contact, stall for time, and cooperate. If you are threatened by a weapon, freeze in place. Never try to grab a weapon.

- Attempts for escape from a confrontation should be to the closest secure area where you can quickly contact others and get help.

- Telephoned threats, like any threat, must be taken seriously. Keep the person talking while you signal a co-worker to get on an extension. Write down what the caller says, and ask for details about when, where, or why the threatened violence will take place. Try to get the caller's name or location. Write down as much detail as you can about the caller—accent, pitch of voice, background noise, etc. Immediately report the call to your supervisors or the police.

After Violence Occurs

Any violent act upsets people. Think ahead to have a plan on how to react after a violent incident. Some items that your organization will want to be prepared for are:

- Getting medical attention for anyone who was hurt during the incident.
- Reporting the incident to supervisors and/or the police.
- Securing the area so evidence is not disturbed.
- Identifying witnesses and interviewing them to get detailed notes on what happened.
- Analyze what happened and make plans to prevent it from being repeated.
- Conducting employee assistance counseling or debriefing sessions to help employees reduce stress and fear so they can better understand and handle the situation.

Work at Working Safely

1. Follow your facility's security guidelines.
2. Report suspicious activity. Don't take matters into your own hands.
3. Take steps for personal safety when you're alone or at risk.
4. Stay calm in a violent situation.
5. After a violent incident, talk about it to reduce stress and fear.

WELDING, CUTTING, AND BRAZING: AVOIDING THE 'TRIPLE THREAT'

Gas fumes, radiation, and shock are very real hazards that you face on the job as a welder. Think about it — a welding arc is hot enough to melt steel, and the light it emits is literally blinding. It generates toxic fumes that are composed of microscopic particles of molten metal. Sparks and molten slag thrown by the arc can fly up to 35 feet and can cause fires and explosions.

The Bureau of Labor Statistics (1997) reported that during 1996 26,100 welders and cutters suffered injuries related to their work. Are you doomed to be injured if you are a welder? The job can be safe if you take the proper precautions and follow safe work practices.

Where Are the Regulations?

The Occupational Safety and Health Administration (OSHA) has developed rules governing welding, cutting, and brazing. These regulations are found at 29 CFR 1910.252 through .255 in Subpart Q. Your employer must train you to operate your welding equipment safely and to understand the welding process so that you perform your welding tasks safely.

What Are the Hazards?

Whenever welding, cutting or brazing occurs, everyone involved in the operation must take precautions to prevent fires, explosios, or personal injuries.

Even in metal cutting or repair jobs that are considered routine, always follow established safety procedures, and resist the temptation to take short-cuts.

There are three basic types of welding operations:

- **Oxygen-fuel gas welding** joins metal parts by generating extremely high heat during combustion.
- **Resistance welding** joins metals by generating heat through resistance created to the flow of electric current.
- **Arc welding** joins or cuts metal parts by heat generated from an electric arc that extends between the welding electrode and the electrode placed on the equipment being welded.

Welding hazards vary, depending on the facility, equipment, number of workers present, and the job at hand. These common dangers are associated with welding:

- Eyes and skin can be damaged from continued or repeated exposure to ultraviolet and infrared rays produced by electric arcs and gas flames (PPE required).
- Closed containers that once held flammables or combustibles can explode under high heat (proper cleaning and purging procedures must be followed before hot-work is started).
- Toxic gases, fumes, and dust may be released during welding and cutting operations.

- Welding or cutting near combustible or flammable materials creates a fire hazard (use proper fire protection; issue hot-work permits for work in hazardous areas).
- Metal splatter and electric shock cause injuries.

Identify the welding hazards specific to **your** workplace.

Compressed Gas Cylinders

A primary danger for oxygen-fuel gas welding operations stems from welding with compressed gas cylinders (CGCs) containing oxygen and acetylene. If CGCs are damaged, gas can escape with great force and the vessel itself can explode, injuring people and damaging property nearby.

One example of this danger is called "rocketing." Rocketing occurs when a CGC ruptures or is damaged. The cylinder can then act like a rocket and break through concrete walls or travel through open spaces.

Look for these danger signals when handling CGCs:

- Leaking (you may be able to hear or smell escaping gas).

WELDING 233

- Corrosion.
- Cracks or burn marks.
- Contaminated valves.
- Worn or corroded hoses.
- Broken gauges or regulators.

In addition, some compressed gases are flammable. This chapter looks at the hazards of both non-flammable and flammable compressed gases.

Non-flammable Compressed Gases

Non-flammable compressed gases do not catch fire easily or burn quickly. However, some will burn and many possess other dangers. The cylinder label and MSDS will tell you about toxic properties and physical hazards. Ammonia, argon, carbon dioxide, nitrogen, oxygen, chlorine, and nitrous oxide are all non-flammable compressed gases. These gases may:

- Cause dizziness, unconsciousness, or suffocation.
- Explode or accelerate fires if mishandled or exposed to heat.
- Be harmful or toxic if inhaled.
- Irritate eyes, nose, throat, and lungs.

Flammable Compressed Gases

Flammable compressed gases have dangers besides those of high pressure. These gases can easily catch fire and burn rapidly. Flammable compressed gases include acetylene, hydrogen, natural gas, and propane.

Flammable compressed gases have the same dangers as non-flammable compressed gases, as well as:

- Ignition from heat, sparks, flames.
- Flash back if vapors travel to ignition source.

Follow Safe Welding Practices

Welding, cutting, and brazing operations are done all the time in industrial plants. There's always something that needs to be repaired, constructed, or taken down. Your organization might have a single portable welding unit to do an occasional spot welding task, or it may have large electric welders to use in daily production. Although these tasks are common, they present serious dangers which can be avoided.

Working safely during welding operations will vary greatly depending on the operations taking place. Here are a few practices for working safely that apply in many situations.

- When working above ground or floor level, use a platform with toeboards and standard railings or safety harness and life line. Also protect workers from stray sparks or slag in the area **below** an elevated surface where welding is taking place.
- Aim the welding torch away from cement or stone surfaces. Moisture within these materials could cause them to explode when they reach a certain temperature.
- When finished welding or cutting, warn other workers of hot metal by marking or putting up a sign. Keep floors clean by putting electrode or rod stubs in an appropriate container. Keep floors clear of tripping hazards; store tools safely.
- Never use bare conductors, damaged regulators, torches, electrode holders, or other defective equipment.
- Do not arc or resistance weld while standing on damp surfaces.

Compressed Gas Cylinders

If you handle CGCs properly, you can safely perform many welding tasks. The following procedures will reduce hazards of handling CGCs:

- Identify a gas and its dangers before you use it. You can find this information on labels, MSDSs and cylinder markings. If you don't know what's in a cylinder, don't use it.
- When accepting an acetylene delivery, make sure it arrived upright. Don't accept CGCs of acetylene that arrive in a horizontal position. Such CGCs can explode.
- Make sure valves, hoses, connectors, and regulators are in good condition.
- Check to see if regulators, hoses, and gauges can be used with different gases. Assume they cannot.

- Never open valves until regulators are drained of gas and pressure-adjusting devices are released. When opening CGCs, point outlets away from people and sources of ignition, such as sparks or flames. Open valves slowly. On valves without hand wheels, use only supplier-recommended wrenches. On valves with hand wheels, never use wrenches. Never hammer a hand wheel to open or close a valve.
- When in storage, keep the steel protective cap screwed on. This step reduces the chance that a blow to the valve will allow gas to escape.

- Minimize banging and clanking of CGCs.
- Don't let cylinders fall or have things fall on them.
- Keep CGCs secured and upright. (But never secure CGCs to conduit carrying electrical wiring.)
- When empty, close and return CGCs. Empty CGCs must be marked "MT" or "Empty." Empty acetylene CGCs **must** be so labeled.

WELDING

Here are several things to remember when moving CGCs:

- Don't drop or bang CGCs.
- Don't roll, drag, or slide CGCs.
- You may carefully roll CGCs along the bottom rim for short distances.
- Don't lift CGCs by the cap unless using hand trucks so designed.
- Ropes and chains should only be used if a CGC has special lugs to accommodate this.
- Some CGCs may require special hand trucks.

Cylinder storage also has safety implications. Remember these guidelines when storing CGCs:

- Store cylinders upright.
- Group cylinders by types of gas.
- Store full and empty cylinders apart.
- Store gases so that old stock is removed and used first.
- To keep cylinders from falling over, secure them with chains or cables.
- Make sure fire extinguishers near the storage area are appropriate for gases stored there.
- Store oxygen CGCs at least 20 feet from flammables or combustibles, or separate them by a 5 foot, fire-resistant barrier.

Keep oil and grease away from oxygen CGCs, valves, and hoses. If your hands, gloves, or clothing are oily, do not handle oxygen CGCs. Oxygen and compressed air are not the same thing. Do not use them interchangeably.

Equipment Inspection and Maintenance

It almost goes without saying that welding equipment should be used according to the manufacturer's instructions. You must be familiar with the correct use and limitations of your welding equipment. In addition, routinely inspect and maintain your welding equipment, including welding cylinders. Inspect cylinders regularly to make sure all parts are in good working order, especially manifolds, distribution piping, portable outlet headers, regulators, hose, and hose connections.

Ventilation

Ventilation techniques vary depending on the size and type of the industry you work in. Some large facilities use sophisticated industrial exhaust systems like state-of-the-art electrostatic precipitators. Very often, however, a relatively simple ventilation method like the appropriate use of wall fans will be all that is required to provide good ventilation during welding operations.

Be aware that general ventilation should never be relied on as the only means of protection when air contaminants are toxic. Where ventilation is poor, you may need to use a respirator.

WELDING

Fire Prevention

You must do your welding in places safe from fire hazards. If the welding project itself must take place at a specific location, all fire hazards in the vicinity of a welding or cutting operation must be moved to a safe place before welding may begin.

When neither the object to be welded nor the fire hazards near it can be moved, then guards must be set up to confine heat, sparks, and slag. Under these circumstances, your employer must issue a written authorization, or hot work permit, outlining the conditions under which the welding may occur.

Your employer must designate a worker as a fire watch whenever welding or cutting is performed in locations where other than a minor fire might develop, or any of the following conditions exist:

- Appreciable combustible material, in building construction or contents, closer than 35 feet to the point of operation.
- Appreciable combustibles are more than 35 feet away but are easily ignited by sparks.
- Wall or floor openings within a 35-foot radius expose combustible material in adjacent areas including concealed spaces in walls or floors.
- Combustible materials are adjacent to the opposite side of metal partitions, walls, ceilings, or roofs and are likely to be ignited by conduction or radiation.

Fire watches must:

- Have fire extinguishing equipment readily available and be trained in its use.
- Be familiar with facilities for sounding an alarm in the event of a fire.
- Watch for fires in all exposed areas.
- Try to extinguish fires only when obviously within the capacity of the equipment available, or otherwise sound the alarm.
- Be maintained for at least a half-hour after completion of welding or cutting operations to detect and extinguish possible smoldering fires.

Confined Spaces

Welding or cutting in a confined space presents its own hazards. Follow confined space entry and rescue procedures. In addition:

- Evaluate the space for its limited work area, any hazardous atmosphere, or a slippery floor surface. Evaluate the space for flammability or combustibility hazards and for toxic fumes that could result from the welding process.

- Perform atmospheric testing for oxygen deficiency and for toxic and flammable or combustible gases before and during entry. If the tests show that flammable or combustible gases are present, the space must be ventilated until safe to enter. If the atmosphere is toxic and cannot be cleared through ventilation, appropriate respiratory equipment must be used. All energy sources that could cause employee injury must be disconnected and locked in the "off" position before entry.
- When working in confined spaces wear a safety harness attached to a life line. A similarly equipped helper should tend the life line to observe the welder and initiate emergency rescue procedures.

- If hot-work inside the space is interrupted, special precautions should be implemented. Disconnect power to arc welding or cutting units and remove the electrode from the holder. Turn off torch valves and shut off the gas supply to gas welding or cutting units at a point outside. Remove the torch and hose from the space, if possible. Cylinders for welding operations should never be placed in confined spaces.

Use PPE When Welding

It's important for welders to wear flame-retardant clothing and protective equipment for the eyes, ears, head, and lungs. Necessary protective gear may include the following, depending on the job:

- Aprons—flame resistant (leather or other material that protects against radiated heat and sparks).
- Leggings—leather or similar protection when doing heavy work.
- Safety shoes—ankle length (low cut shoes may catch slag).
- Protection during overhead work—shoulder cape or cover, skull cap made of leather or other protective material, other flame resistant cap worn under helmet.
- Ear protection—ear plugs, and, on very noisy jobs like high velocity plasma torches, ear muffs.
- Head protection—safety helmet or other head gear to protect against sharp or falling objects.
- Eye protection—operators, welders, or helpers should wear goggles, a helmet, and face shield to provide maximum protection for the particular welding or cutting process used. All filter lenses and plates must meet the test for transmission of radiant energy prescribed in ANSI Standard Z87.1, *Practice for Occupational and Educational Eye and Face Protection*.
- Respiratory protection—If gases, dusts, and fumes cannot be maintained below permissible exposure levels (PELs), welders should wear respiratory protective equipment designated by the National Institute of Occupational Safety and Health (NIOSH).

Clothing Preferences

Welders should cover all parts of their bodies to protect against ultraviolet and infrared ray flash burn. Dark clothing works best to reduce reflection under the face shield. Woolen clothing is preferred for arc welders, as it resists deterioration better than cotton.

Wool, leather, or cotton treated to reduce flammability are preferred for welding. Clothing should be thick enough to prevent flash-through burns, be clean, be free of oil or grease, and have sleeves and collars buttoned. Welders should wear pants without cuffs or front pockets that would catch sparks. Pants legs should cover the tops of shoes or boots.

If worn, thermal insulated underwear should be covered by other clothing and not exposed to sparks or other ignition sources. It should be down-filled or waffle weave cotton or wool. Quilted nylon-shell/polyester-filled underwear and synthetic outer wear won't necessarily ignite more easily than cotton, but it melts as it burns, sticking tightly to skin, which can result in a very serious burn that is hard to treat and slow to heal.

Housekeeping Is a Priority

Keeping welding areas free of combustibles is extremely important in avoiding fires. Never throw used electrodes or rod stubs on the floor—they could ignite a fire. Proper storage of compressed gas cylinders is also an element of fire prevention. As always, store tools in the appropriate place.

Work at Working Safely

Getting the job done safely should always be your first concern. In all welding operations, take time to evaluate the job and implement appropriate safety precautions. This step will not only prevent equipment and machine damage, but it will reduce the risk of an accident that could injure you or a co-worker.

CHAPTER REVIEWS

How To Use Chapter Reviews

The following chapter reviews are to be used at the end of each chapter, to review key points and main ideas. Fill out the review when instructed to do so by the trainer. The reviews are perforated so that you can tear them out easily and turn them in if requested to do so by the trainer.

Each review question is in the form of multiple choice, with four answers to choose from. Only one answer is correct for each question. Circle or underline the letter next to the correct answer, as instructed to do so by your trainer.

Employee _____
Instructor _____
Date _____
Location _____

CONFINED SPACE ENTRY REVIEW

1. A confined space:
    a. Is large enough for an employee to enter.
    b. Has restricted means of entry or exit.
    c. Is not designed for continuous employee occupancy.
    d. All of the above.

2. A permit-required confined space:
    a. Is a confined space that may have a hazardous atmosphere.
    b. Is a confined space that contains materials that could engulf an entrant.
    c. Is a confined space with any other recognized serious hazard.
    d. All of the above.

3. Physical hazards in a permit-required confined space:
    a. Include heat, noise, and mechanical equipment.
    b. Include a lack of oxygen.
    c. Include toxic gases.
    d. None of the above.

4. To eliminate physical hazards from a permit-required confined space:
    a. Open all entry ports.
    b. Force air into the space to ventilate it.
    c. Lock out mechanical equipment, and drain and block pipes or lines.
    d. Provide a two-way radio.

5. The amount of oxygen in a permit-required confined space:
    a. Can be used up as material in the space rusts, rots, ferments, or burns.
    b. Must be at a level of at least 24 percent before anyone can enter the space.
    c. Goes back to normal levels as soon as the space is opened.
    d. Stays at the same level even if argon or nitrogen gas are introduced into the space.

6. Fires and explosions:
    a. Are unlikely in a confined space.
    b. Can be caused by a build-up of flammable vapors that were given off from hazardous materials that used to be in the space.
    c. Can be prevented by wearing personal protective equipment.
    d. Will not happen if the inside of the tank has been emptied.
7. Before entering a permit-required confined space:
    a. Post the permit, turn off the power, and test the air.
    b. Give the permit to your supervisor, lockout mechanical equipment, test the air, and have a rescue plan.
    c. Post the permit, control any hazardous energy, test the air, purge and ventilate the space, wear personal protective equipment, and have a rescue plan.
    d. Have a stand-by rescue person, test the air, and wear a respirator.
8. The air in a permit-required confined space must be tested:
    a. For oxygen, combustibility, and toxicity.
    b. Before entry and during work procedures.
    c. At all levels.
    d. All of the above.
9. Purging a permit-required confined space:
    a. Removes any water, sediment, or other unwanted materials.
    b. Removes hazardous atmospheres from the space.
    c. Eliminates any need for ventilation.
    d. Both a and b.
10. The attendant must:
    a. Maintain eye contact with the entrant.
    b. Maintain constant communication with the entrant.
    c. Maintain constant tension on the entrant's lifeline.
    d. Rescue the entrant immediately after calling for additional help.

Employee _____
Instructor _____
Date _____
Location _____

ELECTRICAL SAFETY REVIEW

1. Training is required for:
    a. Qualified electricians only.
    b. Those workers who work on or near exposed energized parts.
    c. Both those workers who work on or near exposed energized parts and those who do not.
    d. None of the above.

2. Workers who are designated as "unqualified:"
    a. Do not need any training.
    b. Must be trained in safety-related work practices.
    c. Must not operate electrically powered equipment.
    d. Both a and c.

3. Workers who are designated as "qualified:"
    a. Work on or near exposed energized parts.
    b. Must be trained to distinguish exposed live parts from other parts of electrical equipment.
    c. Must be trained to determine the nominal voltage of exposed live parts.
    d. All of the above.

4. The two basic kinds of electricity are:
    a. Static and dynamic.
    b. Current and conducting.
    c. Current and resistance.
    d. Electron flow and circuit.

5. A complete circuit is made up of:
    a. A generator, electron flow, and voltage source.
    b. A consuming device, a load, and a conductor.
    c. A source of electricity, a conductor, and a consuming device.
    d. All of the above.

REVIEW–ELECTRICAL SAFETY

6. Electromotive force (EMF) exerts the difference in potential that causes electrons to flow through a circuit. EMF is measured in:
   a. Amperes.
   b. Volts.
   c. Current.
   d. Resistance.
7. You can get an electrical shock:
   a. When some part of your body becomes part of an electric circuit.
   b. If you touch both wires of an electric circuit.
   c. If you touch one wire of an energized circuit and ground.
   d. All of the above.
8. Electricity can cause fires:
   a. Because of defective or misused electrical equipment.
   b. When wires are improperly spliced or connected.
   c. When circuits create high amounts of heat due to excessive current and resistance.
   d. All of the above.
9. Protect yourself from electrical hazards by:
   a. Carrying tools by their power cords, substituting higher capacity fuses to prevent faults, and storing tools in electrical panels.
   b. Inspecting insulation on power cords, using equipment protected by fuses or circuit breakers, keeping the covers closed over wiring compartments, using grounded equipment, and wearing specialized personal protective equipment as needed.
   c. Using extension cords whenever possible, pulling fuses instead of locking out during repairs, leaving covers off of electrical panels to make inspection easier, removing the ground prong from plugs so tools can be plugged into any outlet, and wearing standard canvas gloves when working with electricity.
   d. All of the above.
10. When repairs are made to electrically powered equipment:
    a. An authorized employee must do the repairs.
    b. The power only needs to be shut off at the machine's control panel.
    c. The power must be shut off at the switch box, the switch must be locked in the "off" position, and a warning tag must be applied.
    d. Both a and c.

REVIEW–ELECTRICAL SAFETY

Employee _____
Instructor _____
Date _____
Location _____

EMERGENCY RESPONSE REVIEW

1. Planning for chemical emergencies includes:
    a. Having an emergency response team in place.
    b. Making arrangements with an outside contractor to respond to chemical emergencies.
    c. Allowing any employees to respond to any chemical releases.
    d. Either a or b.
2. If your employer follows an Emergency Action Plan for chemical emergencies:
    a. Workers are not allowed to assist in handling the response.
    b. The workers who caused the release are responsible for the response.
    c. The janitorial staff is allowed to assist in handling the response.
    d. Supervisors are responsible for responding to the release.
3. If your employer follows an Emergency Response Plan for chemical emergencies:
    a. All employees are to take shelter in a structurally sound part of the building.
    b. All employees are to evacuate the building immediately.
    c. Personnel with specific training may respond to a chemical spill.
    d. None of the above.
4. Members of an emergency response team should know:
    a. The hazards of the chemicals.
    b. How to use spill equipment.
    c. The location of fire extinguishers and emergency exits, and first-aid rules.
    d. All of the above.
5. The buddy system:
    a. Means that at least one person is available to answer the phone.
    b. Means that members of the emergency response team should never enter a chemical emergency situation alone.

c. Means that the incident commander will always have someone to send for supplies.
   d. Means that at least two people must be assigned to each task in the response effort.

6. Emergency responders who are wearing disposable, hooded, chemical-resistant coveralls, a positive-pressure supplied-air respirator, rubber boots, steel toe safety boots, inner and outer rubber gloves, a hard hat, and are carrying a two-way radio meet personal protection at:
   a. Level A.
   b. Level B.
   c. Level C.
   d. Level D.

7. Emergency responders who are wearing disposable, hooded, chemical-resistant coveralls, an air-purifying respirator, rubber boots, steel toe safety boots, inner and outer rubber gloves, a hard hat, and are carrying a two-way radio meet personal protection at:
   a. Level A.
   b. Level B.
   c. Level C.
   d. Level D.

8. Spill carts and spill stations are equipped with:
   a. Pillows, pads, absorbants, or neutralizers.
   b. Brooms, mops, squeegees, buckets, and salvage drums.
   c. Warning tags, tapes, and barricades.
   d. All of the above.

9. Decontamination:
   a. Should be used on protective equipment and clothing.
   b. Is an important part of the emergency response effort.
   c. Is unnecessary in most response efforts.
   d. Both a and b.

10. Following an emergency, the following agencies may need to be notified:
    a. OSHA, the National Response Center, or the EPA.
    b. OSHA, the CDC, or the Dept. of Public Works.
    c. OSHA, the DOT, or the IRS.
    d. OSHA, the FAA, or HUD.

Employee _____
Instructor _____
Date _____
Location _____

ERGONOMICS REVIEW

1. Ergonomics is:
   a. Finding a person who fits the environment.
   b. Changing the job to make the equipment work easier.
   c. Arranging the environment to fit the person.
   d. Exercising the person so the job becomes easier.

2. Ergonomics can reduce worker stress and injury through the design of:
   a. Work stations.
   b. Lighting.
   c. Safety devices.
   d. All of the above.

3. What ergonomic concerns can contribute to injury?
   a. Exposure to excessive vibration or noise, or work with repetitive twisting, forceful, or flexing motions.
   b. Exposure to bloodborne pathogens.
   c. Working where chemicals are used.
   d. Working on high voltage electrically operated equipment.

4. An ergonomics program needs employee involvement because:
   a. Employees design the work stations.
   b. Employees know the jobs they perform.
   c. Employees specify the materials to be used.
   d. All of the above.

5. Disorders involving the muscles, tendons, ligaments, nerves, joints, bones, or supporting body tissue are:
   a. Cardiovascular disorders.
   b. Neuroglial disorders.
   c. Musculo-skeletal disorders.
   d. Pulmonary disorders.

6. Worker/workplace interactions that may lead to injury can involve:
   a. The physical characteristics of the worker (size, strength, etc.).
   b. Heavy lifting.
   c. Repeated twisting or other repeated motions.
   d. All of the above.

7. Musculo-skeletal and nervous system disorders which are caused or made worse by repetitive motions, forceful exertions, vibration, contact with hard or sharp edges, or sustained or awkward postures are called:
   a. Cumulative trauma disorders.
   b. Carpal tunnel syndrome.
   c. Tarsal tunnel syndrome.
   d. Ophthalmological disorders.

8. Factors that can lead to back injuries include:
   a. Excessive twisting, bending, or reaching.
   b. Handling loads that are too heavy or too big.
   c. Poor physical condition.
   d. All of the above.

9. Designing work stations that allow you enough space, allow you to adjust the height of equipment, and allow you to easily reach materials is an example of:
   a. Administrative controls.
   b. Engineering controls.
   c. Procedural controls.
   d. Equipment controls.

10. Ergonomic considerations for selecting tools include:
    a. Handle size, grip strength required, and force required to use the tool.
    b. Cost and durability of the tool.
    c. Balance and weight of the tool.
    d. Both a and c.

Employee _____
Instructor _____
Date _____
Location _____

FIRE PREVENTION REVIEW

1. The best defense against a fire is:
    a. To prevent a fire from starting in the first place.
    b. To train everyone to use fire fighting equipment.
    c. To confine debris or flammable materials to one designated area.
    d. To have fire drills.
2. For a small fire involving wood, cloth, paper, rubber, or plastics, use a:
    a. Type A, or type ABC extinguisher.
    b. Type B, or type ABC extinguisher.
    c. Type C, or type ABC extinguisher.
    d. Specialized extinguisher designed for the specific hazard.
3. For a small fire involving flammable liquids, gases, or grease, use a:
    a. Type A, or type ABC extinguisher.
    b. Type B, or type ABC extinguisher.
    c. Type C, or type ABC extinguisher.
    d. Specialized extinguisher designed for the specific hazard.
4. For a small electrical fire, use a:
    a. Type A, or type ABC extinguisher.
    b. Type B, or type ABC extinguisher.
    c. Type C, or type ABC extinguisher.
    d. Specialized extinguisher designed for the specific hazard.
5. For a fire involving combustible metals such as magnesium, titanium, zirconium, or sodium, use a:
    a. Type A, or type ABC extinguisher.
    b. Type B, or type ABC extinguisher.
    c. Type C, or type ABC extinguisher.
    d. Specialized extinguisher designed for the specific hazard.

6. In order to safely use a fire extinguisher:
   a. You must be trained in how to use it.
   b. You must have the right type of fire extinguisher.
   c. The fire must be small and tame enough to be extinguished with a hand-held extinguisher.
   d. All of the above.

7. "PASS" stands for:
   a. Prevent fires, assess the damage, study fire safety, and store materials properly.
   b. Phone the fire department, arrange for evacuation, stay away from the fire, and smother the flames.
   c. Pull the pin, aim at the base of the fire, squeeze the handle, and sweep at the base of the fire with the extinguishing agent.
   d. Prepare for an emergency, advise the fire department, send for help, and stand clear of the flames.

8. It is important to stay low and crawl to an exit if your building is on fire because:
   a. It would be impossible to see your exit if you were walking.
   b. You have more traction if you crawl.
   c. By crawling, you avoid breathing deadly smoke and heat which rise rapidly in a fire.
   d. The flames always start at the ceiling and work their way down.

9. Safe procedures and equipment for handling flammable liquids can be grouped into four segments:
   a. Storage, transfer, use, and disposal.
   b. Selection, training, containment, and neutralization.
   c. Segregation, supervision, confinement, and decomposition.
   d. None of the above.

10. Compressed gases can also contribute to a fire. To prevent fires involving compressed gases:
    a. Use a hand cart or truck specially designed to move cylinders.
    b. Keep oxygen cylinders away from any oil, grease, or flammable liquids.
    c. Carefully inspect all fittings and connections.
    d. All of the above.

Employee _____
Instructor _____
Date _____
Location _____

FIRST AID/BLOODBORNE PATHOGENS REVIEW

1. What is OSHA's regulation on bloodborne pathogens designed to do?
   a. Limit the number of people who can give first aid.
   b. Provide a set of practices to follow that will protect people from infections caused by germs carried in blood.
   c. Eliminate exposure to bacterial infections.
   d. Prevent patients from being infected by health care personnel.
2. Hepatitis B virus (HBV) and human immunodeficiency virus (HIV) are:
   a. Microorganisms present in human blood that can cause disease in humans.
   b. Bloodborne pathogens.
   c. Diseases.
   d. Both a and b.
3. Universal precautions are:
   a. An approach to infection control where all human blood and certain body fluids are treated as if they were infectious.
   b. Steps that a patient should take when being examined by a doctor.
   c. An approach to infection control where precautions are taken when handling blood that is known to be infected.
   d. Giving vaccinations to everyone who is suspected of being infected.
4. "Good Samaritans:"
   a. Are people who may volunteer to give first aid, even though it isn't part of their job duties.
   b. May want to know about, and follow, exposure control methods.
   c. Are health care workers.
   d. Both a and b.
5. If your work could potentially expose you to bloodborne pathogens, you must be trained in:
   a. First aid techniques, cardiopulmonary resuscitation, how to give oxygen, and how to apply direct pressure to a wound.
   b. How to mix a disinfectant solution.

  c. Bloodborne diseases and how they are spread, control measures, personal protective equipment, Hepatitis B vaccine, exposure evaluation and follow-up, response to emergencies involving blood, warning signs and labels, and your employer's exposure control plan.
  d. All of the above.
6. A document that identifies job classifications and tasks that involve exposure to blood and other potentially infectious materials; outlines engineering and workplace controls, personal protective equipment, and housekeeping procedures; and describes procedures for evaluating exposure incidents is:
  a. An exposure control plan.
  b. Required when employees may be exposed to bloodborne pathogens.
  c. Something that affected employees should read.
  d. All of the above.
7. Autoclaves and containers for used sharps are examples of:
  a. Engineering controls.
  b. Work practices.
  c. Antiseptic equipment.
  d. Personal protective equipment.
8. You should wash your hands with soap and water:
  a. After removing gloves.
  b. After any hand contact with blood or potentially infectious fluids.
  c. Even if you've already used an antiseptic cleanser.
  d. All of the above.
9. Proper containers for used sharps:
  a. Must be made of metal.
  b. Must be puncture resistant, leakproof, and labeled.
  c. Must be at located at least five feet above the ground.
  d. Must be locked at all times.
10. A specific eye, mouth, other mucous membrane, non-intact skin, or parenteral contact with blood or other potentially infectious material that results from the performance of job duties:
  a. Is an exposure incident, and the employee must be offered a confidential medical evaluation.
  b. Requires the employee to have a blood transfusion.
  c. Requires the employee to receive another series of Hepatitis B vaccine.
  d. None of the above.

Employee _____ _____
Instructor _____
Date _____
Location _____

FORKLIFT SAFETY REVIEW

1. Forklifts are different than cars because:
    a. They weigh much more.
    b. The are usually steered by the rear wheels.
    c. They have a three-point suspension system.
    d. All of the above.

2. Why is it important to avoid quick accelerating or braking or turning a corner too fast while driving a forklift?
    a. The center of gravity moves as the truck and the load are moved, and a sudden shift in the center of gravity can make the truck more likely to tip over.
    b. Sudden movements can cause the truck to stall.
    c. Sudden movements cause excessive wear to the truck's steering and braking mechanisms.
    d. Both a and b.

3. Information on the forklift's weight and capacity can be found:
    a. On the truck's mast.
    b. Under the forks.
    c. On the truck's nameplate.
    d. On the control panel.

4. A forklift can be operated by:
    a. Anyone with a valid driver's license.
    b. Anyone with good peripheral vision.
    c. Only the assigned driver or drivers.
    d. Any supervisor.

5. While driving a forklift:
    a. Always face forward.
    b. Be aware of overhead clearances.
    c. Rest your arm outside of the cab in case you want to use a hand signal.
    d. All of the above.

6. When you pick up a load:
   a. Make sure the weight of the load is within the capacity of the truck.
   b. Make sure the load is balanced and secure.
   c. Insert the forks all the way into a pallet, and tilt the mast back to stabilize the load.
   d. All of the above.
7. When you deposit a load:
   a. Stop the forklift completely before raising the load.
   b. Tilt the load forward as soon as it is raised to the correct height.
   c. Stand under the raised load to see if it is properly lined up.
   d. Stack the load, even if it doesn't quite fit all the way onto the rack.
8. When you get off of a forklift for a few moments and you stay within 25 feet of it and the forklift is always within view, you must:
   a. Lower the forks, put the controls in neutral, set the brakes, and shut off the power.
   b. Leave the load raised, put the controls in neutral, and set the brakes.
   c. Lower the forks, put the controls in neutral, and set the brakes. You can leave the power on.
   d. Lower the forks and put the controls in neutral. You do not need to set the brakes.
9. When you inspect the forklift check the:
   a. Hydraulics.
   b. Brakes.
   c. Steering.
   d. All of the above.
10. Travel in reverse:
    a. When you are carrying a load up a hill.
    b. When you can't see over a load in front.
    c. At all times.
    d. Never.

Employee _____
Instructor _____
Date _____
Location _____

HAND TOOLS REVIEW

1. Using hand tools can cause:
    a. Eye injuries.
    b. Cuts, puncture wounds, broken bones.
    c. Minor scrapes, cuts, or bruises.
    d. All of the above.

2. Prevent injuries by:
    a. Relying on your experience to protect you.
    b. Wearing personal protective equipment.
    c. Using clamps to hold your work securely.
    d. Both b and c.

3. When using tools around electrical parts:
    a. Use tools made from non-sparking alloys.
    b. Use tools with handles that are electrically insulated.
    c. Use tools with cushioned grips.
    d. Use conductive tools.

4. Knives with sharp blades:
    a. Need less pressure to cut.
    b. Need more pressure to cut.
    c. Have a greater chance of slipping.
    d. Are more easily damaged.

5. Screwdrivers:
    a. Are the "universal" tool.
    b. Are easily modified to fit various types of screws.
    c. Need to fit the screw.
    d. Still work fine if the tip is rounded.

6. Hammers and mallets:
   a. Have different designs for different jobs.
   b. Are dangerous to use if the handle is loose.
   c. Can damage other tools.
   d. All of the above.
7. Pliers:
   a. Easily grip nuts or bolts so a wrench isn't always needed.
   b. Are designed for bending or pulling material.
   c. Maintain their grip even if they're bent or the gripping face has gouges.
   d. None of the above.
8. Open-ended wrenches:
   a. Should be pulled, with the open end facing you.
   b. Should be used when a heavy pull is needed.
   c. Can be struck to apply more force.
   d. Work better if a cheater bar is placed over the handle to apply more torque.
9. Wood saws:
   a. With fine teeth should be used for rough cuts in green wood.
   b. Should be cleaned and lightly oiled after use.
   c. Have different designs for different uses.
   d. Both b and c.
10. Damaged tools:
    a. Should be taken out of use and repaired or replaced.
    b. Should be returned to the tool chest until someone has time to fix them.
    c. Can still be used if you don't apply too much force to them.
    d. Should be immediately thrown away.

Employee _____
Instructor _____
Date _____
Location _____

HAZARD COMMUNICATION REVIEW

1. OSHA's regulation on hazard communication says that:
   a. Only certain chemicals can be used.
   b. Only chemicals that are used in large quantities need to be evaluated for their hazards.
   c. All chemicals must be evaluated for their hazards, and all information relating to these hazards must be made available to workers.
   d. Only chemical engineers can be allowed to handle large quantities of chemicals.

2. Flammability and reactivity are examples of:
   a. Physical hazards.
   b. Health hazards.
   c. Pyrophoric hazards.
   d. Stability hazards.

3. Examples of health hazards related to chemical use are:
   a. Oxidation.
   b. Carcinogenicity, toxicity, and sensitization.
   c. Water-reactivity.
   d. None of the above.

4. Material safety data sheets (MSDS) must contain:
   a. The identity of the chemical and the name, address, and phone number of the manufacturer, importer, or other responsible party.
   b. The physical and chemical characteristics and the physical and health hazards.
   c. Control measures and precautions for safe handling and use.
   d. All of the above.

5. Copies of MSDS must be:
   a. Kept where you can use them during your workshift.
   b. Kept at a central location when employees must travel between workplaces during the day.
   c. Attached to each container of chemicals.
   d. Both a and b.

REVIEW–HAZCOM

6. Containers of hazardous chemicals must be labeled with:
   a. The identity of the chemical, hazard warnings, and the name and address of the chemical manufacturer, importer, or other responsible party.
   b. The phone number of the chemical manufacturer, importer, or other responsible party.
   c. The identity of the chemical and first aid procedures.
   d. The identity of the chemical, hazard warnings, and required personal protective equipment.

7. A written hazard communication program:
   a. Must include a list of the hazardous chemicals known to be present in your workplace.
   b. Must be posted on a bulletin board.
   c. Must detail how your employer meets the requirements for labels, MSDSs, and employee information and training.
   d. Both a and c.

8. Employees are required to be trained:
   a. Annually.
   b. At the time of their initial assignment and whenever a new hazard is introduced into the workplace.
   c. About how to package chemicals for transportation.
   d. About how to classify waste chemicals as hazardous waste.

9. You must be informed about:
   a. The requirements of the standard.
   b. The amounts of each type of chemical in the workplace.
   c. Any operations in the work area where hazardous chemicals are present.
   d. Both a and c.

10. Employee training must include:
    a. Measures you can take to protect yourself.
    b. The physical and health hazards of the chemicals in the work area.
    c. Details of your employer's hazard communication program.
    d. All of the above.

Employee _____
Instructor _____
Date _____
Location _____

LIFTING TECHNIQUES REVIEW

1. The most common cause of back pain and injury is:
    a. Falling.
    b. Sports activities.
    c. Improper lifting.
    d. Dancing.

2. When a disk presses on a nerve, this is:
    a. A strain or sprain.
    b. A ruptured or slipped disk.
    c. The result of chronic tension or stress.
    d. A kidney or prostate disorder.

3. Poor posture:
    a. Is another factor that can contribute to back injuries.
    b. Puts a strain on your back muscles.
    c. May bend the spine into awkward positions.
    d. All of the above.

4. Other factors that can contribute to back injuries are:
    a. Being overweight.
    b. Infrequent exercise.
    c. Following a proper diet.
    d. Both a and b.

5. Relatively minor strains over time or repeated injuries:
    a. Heal rapidly.
    b. Can accumulate to result in a more serious injury.
    c. Strengthen the back muscles.
    d. Both a and c.

6. You can size up a load before you lift it by:
   a. Testing the weight by lifting at one of the corners.
   b. Carrying the load to a scale to weigh it.
   c. Asking someone else if they think it is too heavy for you to lift.
   d. None of the above.

7. Bend your knees when you pick up and set down a load:
   a. To keep the load as low as possible.
   b. To keep from dropping the load on your feet if it is too heavy.
   c. To allow your stronger leg muscles to do the work.
   d. To get the best grip with your hands.

8. Tips for planning ahead could include:
   a. Making sure your path is free of obstacles before you carry a load.
   b. Putting objects on racks or pallets when they go into storage so they will be easier to pick up later.
   c. Breaking up heavy loads into smaller parts if you can.
   d. All of the above.

9. If back pain is accompanied by weakness or numbness in your lower limbs:
   a. You should see a doctor.
   b. You should rest until the pain goes away.
   c. You should drink plenty of fluids.
   d. You should take up jogging.

10. If you must stand for long periods of time:
    a. Keep your weight evenly balanced.
    b. Change foot positions often.
    c. Avoid leaning to one side.
    d. All of the above.

Employee _____
Instructor _____
Date _____
Location _____

LOCKOUT/TAGOUT REVIEW

1. Lockout is the process of:
    a. Preventing the flow of energy from a power source to a piece of equipment, and keeping it from operating.
    b. Warning equipment operators that they are in danger while operating energized equipment.
    c. Preventing the flow of energy from a utility to the facility.
    d. Preventing mechanics from shutting down energized equipment.

2. Lockout locks:
    a. Can be used to lock tool boxes, storage sheds, etc.
    b. Can be used only for lockout purposes.
    c. Are provided by the employer.
    d. Both b and c.

3. Tagout:
    a. Physically restrains a person from operating a power source.
    b. Serves as a warning.
    c. Must be worn like a badge by the employee who is doing the repairs.
    d. Must be applied using specialized tagout tools.

4. Lockout is required:
    a. During servicing and maintenance where unexpected energization or start-up of equipment could harm employees.
    b. During repair and replacement work, and during renovation work.
    c. During modifications or other adjustments to powered equipment.
    d. All of the above.

5. Those employees who physically lock or tag out equipment for servicing or maintenance:
    a. Are affected employees.
    b. Are authorized employees.
    c. Are always the people who normally operate the equipment.
    d. Do not need training.

REVIEW–LOCKOUT

6. Those workers whose job requires them to operate equipment subject to lockout/tagout:
   a. Are affected employees.
   b. Are authorized employees.
   c. Work in areas where lockout/tagout is used.
   d. Both a and c.

7. Instructions on how to place, remove, and transfer locks and who is responsible for them are part of:
   a. Preparing for a shutdown.
   b. The lockout/tagout procedure.
   c. The notification to affected employees.
   d. Both a and c.

8. When authorized employees perform a shutdown, they must:
   a. Notify all affected employees that they are starting a lock-out procedure.
   b. Locate all energy sources that power the equipment.
   c. Follow the procedures to shut down each machine.
   d. All of the above.

9. If several people are needed to work on a piece of equipment:
   a. Their supervisor is the only one who needs to apply a lock.
   b. The first worker to arrive to the job is the one who applies the locks.
   c. Each worker must apply his or her own lock.
   d. The person who will be the last one to finish the job is the one who applies the locks.

10. Authorized employees must be trained:
    a. To recognize hazardous energy sources, and to know the type and magnitude of the hazardous energy sources.
    b. To perform the lockout/tagout procedure.
    c. When an audit shows deficiencies with the procedure.
    d. ~~All of the above.~~

Employee _____
Instructor _____
Date _____
Location _____

MACHINE GUARDING REVIEW

1. Any machine part, function, or process which may cause injury:
   a. Is required to be removed from the machine.
   b. Is required to be guarded.
   c. Is required to be automated.
   d. Is required to be photographed.
2. The point of operation:
   a. Occurs when the machine's power is activated.
   b. Includes written operating instructions and lockout procedures for the equipment.
   c. Is the point where work is performed on the material (cutting, shaping, etc.).
   d. Never needs to be guarded.
3. Power transmission apparatus:
   a. Includes flywheels, pulleys, belts, chains, gears, etc.
   b. Are components of the mechanical system which transmit energy to the part of the machine doing the work.
   c. Consists of the wiring and switches that provide electricity to the machine.
   d. Both a and b.
4. Hazardous mechanical motions and actions:
   a. Should be recognized so you can protect yourself from their dangers.
   b. Include: rotating, reciprocating, and transverse motions.
   c. Include: cutting, punching, shearing, and bending actions.
   d. All of the above.
5. Guards must:
   a. Prevent contact, be secure, protect from falling objects, create no new hazards, and create no interference.
   b. Prevent contact, be secure, create no new hazards, and create no interference.
   c. Prevent contact, be secure, create no new hazards, create no interference, and be interchangeable.
   d. Prevent contact, be secure, protect from falling objects, create no new hazards, create no interference, and be electrically operated.

6. A fixed guard:
    a. Is the best protection for power transmission apparatus.
    b. Is a barrier which prevents access to danger areas.
    c. May be removed without taking any additional precautions, such as lockout/tagout.
    d. Both a and b.
7. Clothing and protective equipment:
    a. Can contribute to the hazards of machine operation because gloves and loose fitting clothing can become caught between rotating parts, jewelry can catch on moving machine parts, and respirator facepieces can limit vision.
    b. Do not contribute to the hazards of machine operation.
    c. Are never required when a machine is guarded.
    d. None of the above.
8. When you use a power saw:
    a. Look away while you cut to protect yourself from flying debris or sawdust.
    b. Make sure the material you are cutting does not have nails or other foreign objects.
    c. Concentrate on the cut—ignore any warning shouts or instructions you may hear.
    d. Carefully guide the saw if the blade is loose.
9. When you use a grinding wheel:
    a. Leave the work rest set to wherever the last person who used the machine left it.
    b. Lift the machine guards out of the way if they interfere with how well you can see your work.
    c. Make sure that the wheels are firmly held on their spindles before you start the machine.
    d. Apply heavy pressure when you start grinding so you warm up the wheel more quickly.
10. When you use stationary machinery:
    a. Use all guards and safety devices.
    b. Make adjustments and accessory changes when the machinery is turned off and unplugged.
    c. Don't talk to anyone as you use the machinery.
    d. All of the above.

Employee _____
Instructor _____
Date _____
Location _____

PERSONAL PROTECTIVE EQUIPMENT REVIEW

1. The use of personal protective equipment:
   a. Is the only way that your employer can protect employees.
   b. Completes other measures that your employer takes to create a safe work environment.
   c. Does not require any special attention to what is worn by workers.
   d. Does not protect against any hazards.

2. A workplace assessment:
   a. Determines if hazards necessitate the use of PPE.
   b. Determines that PPE can be used instead of guards or other engineering controls.
   c. Determines the number of employees needed for each operation.
   d. Both a and b.

3. When the workplace assessment is done:
   a. Employees should be ready to explain each step of their jobs.
   b. Employees should point out potential job hazards.
   c. Situations where PPE is currently used will be noted.
   d. All of the above.

4. Sources of high temperatures, or chemical and dust exposures are:
   a. Examples of mechanical hazards.
   b. Examples of engineering controls.
   c. Examples of hazards that the employer looks for when doing the hazard assessment.
   d. Examples of hazards that can't be controlled by using PPE.

5. Conducting a hazard assessment allows the employer:
   a. To point out areas of high accident and injury rates.
   b. To identify tools or equipment that need to be repaired or replaced.
   c. To decide if processes or work practices need to be changed.
   d. All of the above.

6. When there is a hazardous situation in your workplace:
   a. It is entirely up to the employer to identify it through conducting a hazard assessment.
   b. Employees have a responsibility to report the hazard.
   c. Employees must do what they can to fix the hazard without reporting it.
   d. A supervisor will notice it eventually.

7. Employees should:
   a. Identify all potential hazards before beginning a task.
   b. Go ahead with the job, even if they are unsure about something.
   c. Take chances and ignore precautions if they are in a hurry to finish.
   d. None of the above.

8. Each employee needs to:
   a. Be aware of potential problems and know what to do about them if they happen.
   b. Know the hazard reporting procedures.
   c. Report hazards as soon as he or she becomes aware of them.
   d. All of the above.

9. The OSH Act:
   a. Gives you the right to safety and health on the job.
   b. Guarantees that right without fear of punishment or reprisal from your employer.
   c. Is enforced by the House of Representatives.
   d. Both a and b.

10. The goal of hazard reporting is to make the workplace safer for everyone. To accomplish this:
    a. Comply with the OSHA standards only when it is convenient.
    b. Report any job-related injury or illness promptly, and seek recommended treatment.
    c. Ignore your employer's safety and health rules when you think they are unnecessary.
    d. Avoid using personal protective equipment on the job, even when it is required.

Employee _____
Instructor _____
Date _____
Location _____

EYE PROTECTION REVIEW

1. Training in eye protection:
   a. Is not required.
   b. Must cover when eye protection is necessary, what types of eye protection are necessary, how to use it, its limitations, and how to take care of it.
   c. Is required before you can be allowed to perform work that requires its use.
   d. Both b and c.

2. "You can walk with a wooden leg, you can chew with false teeth:
   a. And you look silly in safety glasses."
   b. But goggles mess up your hair."
   c. But you can't see with a glass eye."
   d. And you can wear a toupee."

3. The main cause of job-related eye injuries is:
   a. Objects striking a worker's eye.
   b. Dusty conditions.
   c. Poor lighting.
   d. Eye strain.

4. Most workers who suffered eye injuries:
   a. Were wearing safety glasses.
   b. Were not wearing eye protection.
   c. Were wearing goggles.
   d. Were wearing safety glasses and a face shield.

5. Workers who are chiseling, metal working, or hammering risk eye injuries caused by:
   a. Dusts or powders.
   b. Thermal radiation.
   c. Flying particles.
   d. Splashing chemicals.

REVIEW–EYE

6. Steps to take to reduce the risk for eye injuries include:
   a. Reduce the occurrence of foreign objects.
   b. Install equipment guards.
   c. Provide PPE.
   d. All of the above.
7. Portable screens:
   a. Are an example of a guarding device.
   b. Can protect other workers from welding sparks and radiation.
   c. Prohibit entry into a restricted area.
   d. Both a and b.
8. Protective eye and face equipment:
   a. Must comply with guidelines published by the American National Standards Institute, ANSI.
   b. Must be be marked on the frames and lenses to show it complies with ANSI.
   c. Must be approved by NIOSH.
   d. Both a and b.
9. Goggles:
   a. Do not fit as close to your eyes as safety glasses do.
   b. Can't be worn over prescription glasses.
   c. Provide additional protection from liquid splashes and dust.
   d. Do not protect the sides of your eyes.
10. Face shields:
    a. Provide full-face protection against molten metal and chemical splashes.
    b. Should always be used with goggles or safety glasses to make sure your eyes are protected as well as your face.
    c. Are available to be worn with a hard hat.
    d. All of the above.

Employee _____
Instructor _____
Date _____
Location _____

FALL PROTECTION REVIEW

1. Your employer should take steps to prevent falls:
   a. From scaffolds.
   b. From ladders.
   c. From stairs.
   d. All of the above.
2. Fall protection systems:
   a. Prevent or restrain a worker from falling, or safely stop or arrest a worker who falls.
   b. Restrict employees from access to fall hazards, or provide cushions to soften the landing.
   c. Allow employees to design their own guardrails, or provide medical attention after a worker falls.
   d. None of the above.
3. Conventional fall protection systems include:
   a. Escalators, handrails, and full body harnesses.
   b. Guardrail, safety net, and personal fall arrest systems.
   c. Alarms, hoists, and lumbar support belts.
   d. Elevators, cages, and lifelines.
4. Safety nets:
   a. Absorb impacts.
   b. Must be installed more than 50 feet below the workers.
   c. Have mesh openings at least 8 inches on each side.
   d. Only need to be tested right after they were installed.
5. A personal fall arrest system:
   a. Keeps workers from falling.
   b. Stops an employee safely after a fall.
   c. Uses the "buddy system" to keep workers anchored together.
   d. Requires that employees be retrained if they are not following fall prevention procedures.

6. A body harness:
   a. Distributes the shock wave evenly over the body when the fall is stopped.
   b. Has an attachment ring under the worker's chin.
   c. Should have the anchor point of the lanyard located near the worker's feet.
   d. Has hardware than can withstand a load of no more than 1,000 pounds.
7. A lanyard:
   a. Connects the body harness to a deceleration device, lifeline, or an anchor point.
   b. Can be no longer than six feet.
   c. May be a rope or a strap.
   d. All of the above.
8. A self-retracting lifeline:
   a. Will stop a worker's fall within two feet when the breaking mechanism is activated.
   b. Limits the worker to standing in one place.
   c. Has a breaking mechanism that is activated from a sudden jerk on the cable when the worker falls.
   d. Both a and c.
9. A positioning device system:
   a. Supports the worker on a wall so that he or she can work with both hands.
   b. Is rigged so that the worker will not fall more than 10 feet.
   c. Is a type of safety net system.
   d. Requires a spotter to hold the other end of the lanyard.
10. Before you use fall protection, you should find out:
    a. The equipment's application limits.
    b. Anchoring and tie-off techniques.
    c. Emergency rescue plans and implementation.
    d. All of the above.

Employee _____
Instructor _____
Date _____
Location _____

FOOT PROTECTION REVIEW

1. Training in the use of protective footwear:
    a. Is not required.
    b. Must cover when protective footwear is necessary, what types of protective footwear are necessary, how to use protective footwear, its limitations, and how to take care of protective footwear.
    c. Is required, and you must be able to show that you can use safety shoes properly before you can be allowed to perform work that requires their use.
    d. Both b and c.

2. The primary cause of foot injuries is:
    a. Skin disease.
    b. Slips, trips, and falls.
    c. Sharp or heavy objects falling on the foot.
    d. Extreme cold.

3. Of workers who suffered foot injuries, one study showed that:
    a. All of the injured workers were wearing protective footwear at the time of the injury.
    b. 75 percent of the injured workers were wearing protective footwear at the time of the injury.
    c. 50 percent of the injured workers were wearing protective footwear at the time of the injury.
    d. Less than 25 percent of the injured workers were wearing protective footwear at the time of the injury.

4. Safety shoes:
    a. Must meet American National Standards Institute testing requirements.
    b. Have toe guards made of steel, reinforced plastic or hard rubber.
    c. Are worn to protect workers in many types of general industry.
    d. All of the above.

5. Metatarsal guards:
   a. Extend past the toes to protect the upper part of the foot from impacts.
   b. Do not have to meet ANSI requirements.
   c. Must also be chemical resistant.
   d. All of the above.
6. Conductive shoes:
   a. Are worn when working near open electrical circuits.
   b. Prevent the accumulation of static electricity.
   c. Keep electrostatic discharge from igniting sensitive explosive mixtures.
   d. Both b and c.
7. Electrical hazard shoes:
   a. Are worn when working on live electrical circuits.
   b. Have no exposed metal.
   c. Should be kept dry.
   d. All of the above.
8. Rubber or plastic safety boots:
   a. Are worn when working on live electrical circuits.
   b. Prevent static build-up.
   c. Offer protection against chemicals.
   d. None of the above.
9. Add-on foot protection:
   a. Is not allowed.
   b. Can include metatarsal guards, rubber spats, or puncture resistant inserts.
   c. Is not very useful.
   d. Both b and c.
10. Safety shoes:
    a. Are not available in fashionable styles.
    b. Can't use foam latex inner soles to make them more comfortable.
    c. Will not lessen the severity of injuries caused by heavy equipment.
    d. None of the above.

Employee _____
Instructor _____
Date _____
Location _____

HAND PROTECTION REVIEW

1. Hand protection is required when workers are exposed to hazards such as:
    a. Severe cuts, lacerations, or abrasions.
    b. Punctures.
    c. Chemical burns.
    d. All of the above.

2. After training, you must:
    a. Show that you understand the training and can use hand protection properly.
    b. Be able to repair hand protection.
    c. Prevent situations that could create hazards to your hands.
    d. All of the above.

3. Three basic types of hazards that hands are exposed to are:
    a. Lacerations, abrasions, and punctures.
    b. Mechanical hazards, environmental hazards, and irritating substance hazards.
    c. Point of operation hazards, radiation hazards, and reactivity hazards.
    d. Crushing hazards, excessive heat, excessive moisture.

4. The first defense to reduce hand injuries is:
    a. Personal protective equipment.
    b. Engineering controls.
    c. Administrative controls.
    d. Work practice controls.

5. Gloves:
    a. Should be designed to protect against specific hazards.
    b. For electrical lineworkers are color coded to match their level of voltage protection.
    c. Are the most commonly used type of PPE for hands.
    d. All of the above.

6. People who work with sharp knives and saws can protect their hands with:
    a. Leather gloves.
    b. Metal mesh gloves.
    c. Leather mitts.
    d. Neoprene gloves.

7. Barrier creams:
    a. Must be applied frequently.
    b. Are helpful when other types of protection can't be worn.
    c. Do not protect against highly corrosive substances.
    d. All of the above.

8. Carpal tunnel syndrome:
    a. Has pain and paralysis as early symptoms.
    b. Is an ergonomic problem related to the tendons connecting the fingers and the elbow.
    c. May require surgery.
    d. Is treated by having the patient perform repetitive, forceful hand and wrist movements.

9. First aid treatment for cuts includes:
    a. Putting the hand under running water and flushing for 10 to 20 minutes.
    b. Applying a tourniquet.
    c. Applying direct pressure.
    d. Keeping the area cool.

10. Injuries to the hands or fingers:
    a. Account for about 12 percent of work-related injuries.
    b. Happen to over 200,000 workers in an average year.
    c. Can be severe.
    d. All of the above.

Employee _____
Instructor _____
Date _____
Location _____

HEAD PROTECTION REVIEW

1. Head injuries:
    a. Are never fatal.
    b. Are only caused by falling objects.
    c. Can cause concussions, trauma, or death.
    d. Only happen in construction work.

2. Head injuries are caused by:
    a. Falling objects.
    b. Flying objects.
    c. Electrical shock.
    d. All of the above.

3. Hard hats are designed to:
    a. Resist the penetration and absorb the shock from a blow.
    b. Resist the penetration and absorb the shock from a blow, and to provide protection from electrical shock and burn.
    c. Keep you safe from cold weather and intense sun.
    d. None of the above.

4. Hard hats lessen injuries:
    a. Because the sweatband helps to keep your eyes clear.
    b. Because the hard outer shell will not crack.
    c. Because the hard outer shell and suspension system work together to absorb impacts.
    d. Because you take fewer risks while wearing a hard hat.

5. Type 1 hard hats:
    a. Do not protect from falling objects.
    b. Do not protect from electrical hazards.
    c. Do not require a suspension system.
    d. Have a full brim.

6. Type 2 hard hats:
    a. Have a full brim.
    b. May have a peak extending over the eyes.
    c. Always require a chin strap.
    d. May not be worn in cold weather.
7. Class A hard hats:
    a. Offer no voltage protection.
    b. Offer limited voltage protection.
    c. Offer the highest voltage protection.
    d. Will not have a brim.
8. Class B hard hats:
    a. Offer no voltage protection.
    b. Offer limited voltage protection.
    c. Offer the highest voltage protection.
    d. Can't be used with ear protection.
9. Class C hard hats:
    a. Offer the most impact resistance.
    b. Offer the most voltage protection.
    c. Offer no voltage protection.
    d. Both a and b.
10. Dents, cracks, and paint on a hard hat:
    a. Do not affect the helmet's performance.
    b. Can limit the protection provided by the helmet.
    c. Are easily repaired.
    d. Mean you are working hard.

Employee _____
Instructor _____
Date _____
Location _____

HEARING CONSERVATION REVIEW

1. The ear works to:
   a. Concentrate sound waves and direct them to the brain.
   b. Change sound waves into nerve impulses which are interpreted by the brain.
   c. Absorb nerve impulses and transform them into sound waves.
   d. Create sound waves and transfer them to the nerves in the spinal cord.

2. Sound intensity:
   a. Is the loudness of sound.
   b. Is measured in decibels.
   c. Both a and b.
   d. None of the above.

3. Sound frequency:
   a. Is the pitch of a sound (high or low).
   b. Is measured in decibels.
   c. Both a and b.
   d. None of the above.

4. Too much noise:
   a. Can make you tired and irritable.
   b. Is linked to high blood pressure, ulcers, headaches, sleeping disorders.
   c. Both a and b.
   d. None of the above.

5. Steps to reduce noise might include:
   a. Placing machinery on metal mountings to reduce vibration.
   b. Using sound-reflecting tiles on floors, walls and ceilings.
   c. Moving noisy machinery to a separate area, or building a sound barrier around it.
   d. All of the above

6. When equipment can't be made to run quieter, and there are still hazardous levels of noise after controls have been put in place:
   a. You have to "get used" to all the noise.
   b. You can protect yourself with hearing protective devices.
   c. You can shut off the machines whenever you are in the area.
   d. You can work outside.

7. The selection of the right hearing protection depends on:
   a. The noise levels and the Noise Reduction Rating of the hearing protector.
   b. The frequency of the noise and whether it is continuous or intermittent.
   c. The fit and comfort of the protective devices.
   d. All of the above.

8. Hearing protectors:
   a. Block out all sound completely.
   b. Overload your hearing with surrounding noises.
   c. Reduce the amount of sound reaching the delicate parts of the ear.
   d. Are designed to interfere with speech and machinery sounds.

9. Earmuffs:
   a. Must always be worn with the headband over the head.
   b. Usually reduce sound levels by 20 to 25 decibels.
   c. Have cups made of durable stainless steel.
   d. Are made up of cups, cushions, the headband, and the chin strap.

10. Audiometric testing:
    a. Is required when noise exposure equals or exceeds an average of 85 dBA over an eight-hour day.
    b. Uses an audiometer to send test sounds through headphones worn by the person being tested.
    c. Produces an audiogram that is a chart of the person's responses to the test sounds.
    d. All of the above.

Employee _____
Instructor _____
Date _____
Location _____

RESPIRATORY PROTECTION REVIEW

1. Exposure limits are:
    a. Safe levels of contaminants that employees can be exposed to each workday.
    b. Used by employers to help decide if respirators are needed.
    c. Areas where employees are denied entry.
    d. Both a and b.

2. The two types of respirators are:
    a. Positive pressure and pressure-demand.
    b. Self-contained and portable.
    c. Air-purifying and atmosphere-supplying.
    d. Negative pressure and air-purifying.

3. Air-purifying respirators:
    a. Can be used where there is a lack of oxygen.
    b. Remove the contaminants from the air as you breath.
    c. Typically have loose-fitting hoods.
    d. None of the above.

4. Atmosphere-supplying respirators:
    a. Provide breathing air from a clean source.
    b. Can be Supplied-Air Respirators or Self-Contained Breathing Apparatus.
    c. Are used by fire fighters.
    d. All of the above.

5. Air-purifying respirator cartridges and filters:
    a. Must be approved for the dust, mist, fume, vapor, or gas that you are exposed to.
    b. Do not need to be changed until you notice a smell or taste inside of the respirator.
    c. Safely remove all types of hazardous substances at any amount of contamination.
    d. Have no limits on how long they will protect you.

REVIEW–RESPIRATORY

6. Before you can be required to wear a respirator:
   a. You will need to fill out a medical questionnaire.
   b. You might need an exam and medical tests.
   c. Your employer will tell the health care professional about your job conditions.
   d. All of the above.

7. Respirators with tight-fitting facepieces require a fit test:
   a. Only if the respirator feels uncomfortable.
   b. Only if you have a beard.
   c. Before you can be required to wear the respirator, with re-testing at least annually.
   d. None of the above.

8. The positive pressure check and negative pressure check are:
   a. Quality control tests for atmosphere-supplying respirators.
   b. Tests that you need to do to check the seal when you put on a tight-fitting facepiece.
   c. Items to look at when you inspect your respirator.
   d. Part of the medical evaluation.

9. Types of respirators that must be inspected monthly are:
   a. Respirators that are kept for emergency use.
   b. Respirators that are issued to one person.
   c. Respirators that are shared for regular use.
   d. Respirators that are used in confined spaces.

10. When you inspect a respirator:
    a. Check the condition of the facepiece, straps, and valves.
    b. Make sure rubber parts are pliable and are not cracked.
    c. Check that all parts are functioning properly.
    d. All of the above.

Employee _____
Instructor _____
Date _____
Location _____

SLIPS, TRIPS, AND FALLS REVIEW

1. Falls:
    a. Are the cause of about 10% of workplace deaths.
    b. Are never fatal.
    c. Usually cause a worker to lose more than 25 work days because of injuries.
    d. Don't cause broken bones or back injuries.

2. Poor friction between your shoes and the walking surface:
    a. Could cause you to slip and fall.
    b. Means your shoes can't "grip" the surface.
    c. Means you lose traction.
    d. All of the above.

3. Momentum:
    a. Involves the size and speed of the moving object.
    b. Is why your fall is harder when you are moving faster.
    c. Increases the friction between two objects.
    d. Both a and b.

4. Slips can be avoided by:
    a. Taking large steps when walking on wet surfaces.
    b. Walking around spills until someone else reports them.
    c. Cleaning up grease from the floor around machinery.
    d. Carefully sliding quickly across smooth, slippery surfaces.

5. When a floor surface is wet frequently:
    a. You can paint it with a gritty, abrasive coating.
    b. You can apply strips of skid-resistant material.
    c. You can use a rubber mat in the slippery area.
    d. All of the above.

6. To avoid tripping:
   a. Walk slowly when you can't see over a load.
   b. Keep aisles or stairs clean and free of clutter.
   c. Hang brightly colored string or ribbons from extension cords so pedestrians will know to step over them.
   d. Cover broken floor boards or pavement with rugs or mats.

7. To avoid slips, trips and falls on stairs:
   a. Grab for the handrail as you fall.
   b. Step over broken stair treads until someone else reports them.
   c. Report burned-out bulbs in stairways.
   d. Only run down the stairs when you are within a few steps of the bottom.

8. Falls from heights:
   a. Do not happen as often as falls from slips or trips at ground level.
   b. Pose a much higher risk of serious injury than falls at ground level.
   c. Can be caused by misuse of ladders or scaffolding.
   d. All of the above.

9. When using a ladder:
   a. Allow two people to climb it at the same time if it would make the job easier.
   b. Check the ladder's condition before climbing.
   c. Stand on the top step of the stepladder, or set the ladder on top of blocks, if it isn't tall enough.
   d. The distance from the wall to the base of the ladder should be the same as the distance from the base of the ladder to where it touches the wall.

10. When using a scaffold:
    a. Check it daily for any safety defects.
    b. Never overload the scaffold with people, equipment, or supplies.
    c. Lock the casters on mobile scaffolding so it will not move while you are working on it.
    d. All of the above.

Employee _____
Instructor _____
Date _____
Location _____

VIOLENCE IN THE WORKPLACE REVIEW

1. What is the second leading cause of all fatalities in the workplace?
    a. Vehicle collisions.
    b. Homicide.
    c. Falls.
    d. Electrocution.

2. Sources of workplace violence are:
    a. Disputes among co-workers.
    b. Robbery and other crimes.
    c. Domestic violence.
    d. All of the above.

3. What are some clues that a situation could become violent?
    a. A person wants to discuss a problem.
    b. A person is making theats.
    c. A person is loitering or a stranger is in a restricted area.
    d. Both b and c.

4. If you need to travel as part of your job, protect yourself by:
    a. Checking in with your co-workers often.
    b. Listening to the radio.
    c. Taking frequent breaks in crowded coffee shops.
    d. None of the above.

5. Having good lighting and lots of windows and mirrors:
    a. Makes it easier for robbers to find the cash drawer.
    b. Makes it more difficult for a co-worker to start an argument.
    c. Makes it easier for security patrols to watch the workplace.
    d. Prevents criminals from entering the workplace.

6. When traveling, avoid becoming a target for violence by:
    a. Approaching your vehicle with your keys ready to use.
    b. Looking in and around your vehicle before getting in.
    c. Locking your doors before you start your vehicle.
    d. All of the above.

7. Employers can reduce the risk for violence by providing:
    a. Video cameras, alarms, and drop safes for cash.
    b. Security guards and metal detectors.
    c. Identification badges and visitor passes.
    d. All of the above.

8. If you witness a potentially violent situation:
    a. Report suspicious activity or threats immediately.
    b. Confront the person and demand that he or she leaves the area.
    c. Follow the suspicious person until the police arrive.
    d. Threaten the violent person.

9. If you receive a threat over the telephone:
    a. It probably is not serious.
    b. Talk to the caller to get as much information as you can, and report the incident.
    c. Transfer the call to your organization's Security Officer.
    d. Hang up.

10. If there has been a violent incident in your workplace:
    a. Go home immediately.
    b. Call the local newspaper and radio station to report the incident.
    c. Get medical help for the injured, report to the police, secure the area, identify and interview witnesses, plan for prevention, provide counseling.
    d. All of the above.

Employee _____
Instructor _____
Date _____
Location _____

WELDING REVIEW

1. What are the three basic types of welding operations?
    a. Brazing, oxy-acetylene welding, and arc welding.
    b. Resistance welding, oxy-acetylene welding, and soldering.
    c. Oxygen-fuel gas welding, resistance welding, and arc welding.
    d. Spot welding, oxygen arc cutting, and arc welding.

2. Some of the hazards associated with welding are:
    a. Eye and skin damage from exposure to ultraviolet and infrared radiation.
    b. Exposure to toxic gases, fumes, and dusts.
    c. Fire and explosion hazards.
    d. All of the above.

3. Signs of damage to compressed gas cylinders and equipment are:
    a. The sound or smell of escaping gas.
    b. Worn hoses.
    c. Broken regulators.
    d. All of the above.

4. Examples of non-flammable gases are:
    a. Carbon dioxide, oxygen, and nitrogen.
    b. Acetylene, propane, and natural gas.
    c. Both a and b.
    d. None of the above.

5. Examples of flammable gases are:
    a. Carbon dioxide, oxygen, and nitrogen.
    b. Acetylene, propane, and natural gas.
    c. Both a and b.
    d. None of the above.

6. Precautions for welding operations include:
   a. Taking down warning signs as soon as you are done welding.
   b. Using shields or tarps to protect workers and equipment below the welding job when you are working above ground level.
   c. Being extra careful when you use damaged regulators or torches.
   d. Putting electrode or rod stubs into your pockets.
7. Precautions for handling compressed gas cylinders include:
   a. Only accepting acetylene cylinders that arrive in a horizontal position.
   b. Making sure valves, hoses, connectors and regulators are in good condition.
   c. Using the same regulators, hoses, and gauges with all different types of gases.
   d. Using any type of wrench or hammer on valves without hand wheels.
8. A fire watch must be assigned when welding operations are done close to combustible materials or where a sizeable fire might develop. Fire watches must:
   a. Be trained to operate fire extinguishing equipment, and have it ready for use.
   b. Be maintained for at least a half-hour after completion of welding or cutting operations.
   c. Try to extinguish fires or sound the alarm as appropriate.
   d. All of the above.
9. When welding in a confined space:
   a. Evaluate the space for combustible or flammable materials before you start.
   b. Follow all confined space entry and rescue procedures.
   c. Monitor the atmosphere for oxygen, flammables, and toxics before and during entry.
   d. All of the above.
10. Welders should wear:
    a. Clothing that covers all parts of the body to protect from welding flash.
    b. Synthetic clothing because it won't melt as it burns.
    c. Pants legs with cuffs.
    d. Open collars that can't be buttoned tight.

REVIEW–WELDING